Enterprise Dynamics Sourcebook

COMPLEX AND ENTERPRISE SYSTEMS ENGINEERING

Series Editors: Paul R. Garvey and Brian E. White
The MITRE Corporation
www.enterprise-systems-engineering.com

C & E S E

Architecture and Principles of Systems Engineering
Charles Dickerson and Dimitri N. Mavris
ISBN: 978-1-4200-7253-2

Designing Complex Systems: Foundations of Design in the Functional Domain
Erik W. Aslaksen
ISBN: 978-1-4200-8753-6

Engineering Mega-Systems: The Challenge of Systems Engineering in the Information Age
Renee Stevens
ISBN: 978-1-4200-7666-0

Enterprise Systems Engineering: Advances in the Theory and Practice
George Rebovich, Jr. and Brian E. White
ISBN: 978-1-4200-7329-4

Leadership in Chaordic Organizations
Beverly G. McCarter and Brian E. White
ISBN: 978-1-4200-7417-8

Model-Oriented Systems Engineering Science: A Unifying Framework for Traditional and Complex Systems
Duane W. Hybertson
ISBN: 978-1-4200-7251-8

FORTHCOMING

Systems Engineering Economics
Ricardo Valerdi
ISBN: 978-1-4398-2577-8
Publication Date: December 2013

RELATED BOOK

Enterprise Dynamics Sourcebook
Kenneth C. Hoffman, Christopher G. Glazner, William J. Bunting, Leonard A. Wojcik, and Anne Cady
ISBN: 978-1-4200-8256-2

AUERBACH PUBLICATIONS

www.auerbach-publications.com
To Order Call: 1-800-272-7737 • Fax: 1-800-374-3401
E-mail: orders@crcpress.com

Enterprise Dynamics Sourcebook

Edited by Kenneth C. Hoffman • Christopher G. Glazner
William J. Bunting • Leonard A. Wojcik • Anne Cady

CRC Press
Taylor & Francis Group
Boca Raton London New York

CRC Press is an imprint of the
Taylor & Francis Group, an **informa** business

AN AUERBACH BOOK

CRC Press
Taylor & Francis Group
6000 Broken Sound Parkway NW, Suite 300
Boca Raton, FL 33487-2742

First issued in paperback 2018

© 2013 by Taylor & Francis Group, LLC
CRC Press is an imprint of Taylor & Francis Group, an Informa business

No claim to original U.S. Government works

ISBN-13: 978-1-4200-8256-2 (hbk)
ISBN-13: 978-1-138-38142-1 (pbk)

Visit the Taylor & Francis Web site at
http://www.taylorandfrancis.com

and the CRC Press Web site at
http://www.crcpress.com

Contents

SECTION I FOUNDATIONS AND CONCEPTUAL FRAMEWORKS

SECTION II ENTERPRISE MODELING APPROACHES AND APPLICATIONS

Acknowledgments

The authors sincerely thank those who contributed case study chapters for this sourcebook. These represent an impressive sampling from over a decade of enterprise planning and analytic experience at MITRE.

The organizing principles of enterprise dynamics as a core enterprise systems engineering discipline stem from MITRE-sponsored research performed over a three-year period from 2005 to 2007. David Lehman, then MITRE's chief engineer, and Anne Cady, then chief engineer of MITRE's Center for Enterprise Modernization and a co-author, actively supported the initiation of the research effort. Kirkor Bozdogan and Joseph Sussman of the Engineering Systems Division at the Massachusetts Institute of Technology collaborated in the research and made invaluable contributions that are evident in the numerous citations of their work.

Rick Sciambi, chief engineer of the MITRE Center for Connected Government, and Glenn Roberts, chief engineer of the Center for Advanced Aviation System Development, provided much appreciated counsel and guidance to the wide range of projects and research efforts represented in this sourcebook.

Many thanks to Iris Fahrer for editing support, and to others at Taylor & Francis for their helpful contributions.

Introduction

The only constant is change, continuing change,
inevitable change that is the dominant factor in society today.

Isaac Asimov

This *Enterprise Dynamics Sourcebook* is the product of The MITRE Corporation's (MITRE) Sponsored Research Program and a range of published applications that pertain to enterprise transformation programs. As a sourcebook it draws on a series of published papers and reports documenting case studies that provide a starting point for a new discipline of enterprise dynamics as a core capability of enterprise systems engineering (ESE). This case study approach captures the diversity of transformation environments and the evolution of methods to deal with this emergent challenge to government and private sector enterprises. Just as fluid dynamics and structural dynamics advance other engineering practices, enterprise dynamics advances the practices of systems engineering at the enterprise level and enterprise architecting, enabling the enterprise systems engineering and architecting (ESE/A) process described in Chapter 5, Section 5.8 and drawing on tools and methods described throughout Section I.

Rapidly changing market, technological, and organizational environments pose complex challenges to government and private enterprises that must improve services and transform their processes, organizations, and resource base. Planning and management of such extensive transformation require extended management tools and methods that deal with the dynamics of change. This sourcebook provides a foundation and examples of methods that deal with the emerging complexities of enterprise transformation involving the coordination of policies, organizations, economics, and technology (POET) in operational strategies and processes. In its original formulation, the "O" in POET designated operations; here it is used to designate organizations, thus giving greater emphasis to organizational issues, as the major theme of ESE/A is transformation of operational capabilities, and recognizing the importance of managing organizational change in coordination with other transformation activities. Throughout this sourcebook enterprise dynamics focuses on the mission and business operations of the enterprise.

The concepts discussed in this sourcebook are applicable to both government and large private sector enterprises that involve significant service or regulatory interactions with government entities and must transform to operate effectively in this complex ever-changing environment.

Dynamic phenomena are a very important part of all organizational theories and engineering disciplines and are a central theme of the analytic case studies in this sourcebook. In these enterprise dynamics case studies, we describe and apply frameworks and analytical models to a range of enterprises to improve the understanding of their dynamic elements and interactions. Specific enterprise transformation efforts must involve all of the POET elements over the full life cycle of transformation with a balance appropriate to the specific enterprise. Planning, implementing, and managing the transformed enterprise require multiple skills and methods. Just as dynamic analyses in engineering and the social sciences draw on a wide variety of methods applicable to specific challenges, the applications described here were selected to introduce a wide range of methods that have demonstrated their utility for the analysis of enterprise dynamics.

The sourcebook is directed toward analysts, managers, and decision makers engaged in significant transformation of their enterprises to operate at high performance levels in the delivery of products and services. The concepts apply in the public and private sectors and address the inherent technical, social, economic, and management complexities of enterprise operations and transformation in a global economy. It is written with the belief that qualitative and quantitative analytical methods drawn from systems engineering and management science can inform the managers of enterprise transformation programs and reduce their risks and unacceptably high failure rates.

Working Definition of an Enterprise

An enterprise is a purposeful social, technical, and economic undertaking designed to create value for its stakeholders involving:

- Policies, organizations (people), economics, and technology interacting with each other and their environment in operational processes as a complex system-of-systems to achieve goals
- An organization (single or multiagent, and possibly virtual) created for the undertaking
- A readiness to embark on bold new ventures

The dynamics of an enterprise are highly dependent on the mission or business characteristics (the enterprise landscape) and the perspective of the enterprise required to address management issues and inform decision making (operational, scalar, and temporal perspectives).

The case studies in Section II represent enterprise "purposeful undertakings" at multiple scales ranging from a commercial business strategy, through technology challenges, an operations control center, and agency activities, to national programs for the transformation of healthcare and the U.S. energy system.

Working Definition of Enterprise Transformation

The transformation of an enterprise to a desired future state is a comprehensive, managed change process driven by public (consumer) demands, competitive factors, economic forces, and technological opportunities. The primary transformation objectives are significant performance improvements and the effective coordination of policies, organizations, economics, and technology in operational strategies and processes. Performance improvements include better products and services, cost, and quality. The dynamic and iterative change process and key elements may be portrayed as shown in Figure 0.1.

The agile sense-and-respond enterprise is quick in sensing and responding to external events and trends and is in a constant state of transformation given the pace of socioeconomic change. This transformation requires proactive analysis of the dynamic interactions among policies, organization, economics, technology, and the environment, incorporating inherent risks and uncertainties.

Each of the case studies in Section II represents unique transformation challenges and draws upon applicable models and approaches from those described in Section I.

The initial chapters in Section I describe the scope and complexities of enterprise transformation as well as approaches to planning and managing that transformation, including ESE/A and the supporting discipline of enterprise dynamics. These emergent and comprehensive methods extend systems engineering and architecting into the complex enterprise environment to address the coordinated evolution and transformation of all POET elements of the enterprise to adapt to changing public demands and expectations, and to take advantage of technical and organizational opportunities. This is the most recent stage in the long-term continuous

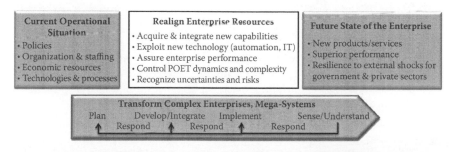

Figure 0.1   Goal: Transformation to an agile sense-and-respond enterprise.

evolution of systems engineering to address issues of increasing complexity, in this case enterprises operating at a national and global scale involving multiple partners and stakeholders, divergent interests, and a high degree of risk and uncertainty.

Section I also presents a selection of analytical tools and methods (models) that deal with specific complexities of transformation addressed through ESE/A. Collectively the tools and methods establish an initial foundation for ESE/A and the supporting discipline of enterprise dynamics, an emergent and comprehensive field that extends systems engineering and architecting into the complex enterprise environment.

Section II presents a series of ESE/A case studies based on published papers and reports on modeling the complex dynamics in specific enterprise-level applications to address critical issues in the acquisition of new operating capabilities. They represent enterprise transformation challenges at a scale where the government role is paramount and, in some of the applications presented, with significant private sector and consumer involvement. The case studies employ a variety of modeling and analytical methods as appropriate to the specific enterprise transformation challenge they are addressing. Examining this variety in the context of the ESE/A process provides valuable and practical guidance for management and problem solving.

Concepts and characteristics of complex systems and megasystems were introduced in an enterprise context by Hybertson (2009), Stevens (2010), and Rebovich and White (2010). This sourcebook supplements those concepts. It describes a discipline of enterprise dynamics incorporated in an ESE/A process to address the planning, design, and management of complex enterprises and megasystems in a very challenging technical, social, and economic environment. Illustrative applications pertaining to a range of enterprise domains are summarized to demonstrate how the process can be tailored to specific enterprise challenges with a focus on mission performance. A top-level summary of the sourcebook chapters is presented below. Each chapter title is followed by key concepts within the chapter.

Section I: Foundations and Conceptual Frameworks

Section I of this sourcebook establishes the foundations of ESE/A and the supporting discipline of enterprise dynamics, an emergent and comprehensive field that extends systems engineering and architecting into the complex enterprise environment. This is the most recent stage in the long-term continuous evolution of systems engineering to apply quantitative and qualitative methods to issues of increasing complexity, in this case enterprises operating at a national and global scale involving multiple partners and stakeholders, divergent interests, and a high degree of risk and uncertainty.

The discipline of ESE/A merges system engineering with management science, enhancing current concepts of enterprise architectures (EA) to deal with increasingly complex challenges faced by government agencies and large commercial enterprises. The emphasis is on describing, analyzing, and evaluating dynamic behaviors of complex techno-socio-economic systems, with all of the POET elements of the enterprise working together for effective operational purposes (enterprise dynamics). Several methods are described for dealing with complexities, risks, and uncertainties to provide a robust adaptive approach to planning and implementing major national and international programs.

Chapter 1: Defining Enterprise and Transformation Challenges

The enterprise definition that best captures the transformation theme and the range of scales addressed is: an entity engaged in a purposeful undertaking that involves internal and external stakeholders, and operates as a complex adaptive system involving policy, organizational, economic, and technology dimensions. This working definition clearly includes complex systems and megasystems and establishes a scope that encompasses a broader boundary that includes all stakeholders rather than just an individual organization. This broad definition can include specific agencies with a complex mission, multiagency initiatives, or virtual organizations that come together for a specific purpose. The transformation challenges that are outlined establish the need for an improved planning and management process that incorporates enterprise dynamics. Effective governance of the enterprise and transformation programs is critical to success and this aspect of a management structure is discussed in Chapter 5 as a wrap-up to Section I. Decision support and resource management for such enterprises can be improved by systems engineering and architecting methods applied at the enterprise level: ESE/A.

Chapter 2: From Systems Engineering to Enterprise Systems Engineering

An historical perspective illuminates the long-term evolution of the field of systems engineering in response to public and private sector challenges and leads up to a foundation for the enterprise systems engineering methodology as described in a following chapter. This evolution involves increasingly complex challenges and evolving tools and methods for planning and analysis. Historical examples reveal a tendency toward technical hubris in expecting more than the methods are capable of delivering, a warning that the enterprise systems engineer must draw upon management and organizational theory and approaches to decision making under uncertainty, in addition to other analytical methods that serve to describe and analyze complex systems.

Chapter 3: Foundations of Enterprise Systems Engineering and Architecting

The current state of evolution in enterprise systems engineering and enterprise architecting is reviewed in this chapter and is extended to portray an integrated approach to ESE/A drawing on concepts of enterprise dynamics. Managing and enhancing the performance of public and private enterprise activities are the subject of architecting, modeling, and simulation applying the disciplines of management science, probability theory, organizational theory, systems engineering, and socioeconomics. Each discipline addresses specific dynamic elements of the enterprise from a unique perspective and contributes a robust foundation for ESE/A that addresses the complexity and dynamics of the interacting POET elements of the enterprise. This field has been fertile ground for research and applications over the past few years and is sufficiently mature to identify foundational concepts.

Chapter 4: Enterprise Dynamics Methods and Models

This chapter summarizes a diverse set of enterprise dynamics methods and models that have been applied in the ESE/A applications summarized in this sourcebook. They constitute a representative sampling of the differing perspectives across disciplines applied to a range of enterprise challenges, and comprise a new discipline of enterprise dynamics (or perhaps a subdiscipline in the larger ESE/A context). Emphasis is placed on characterizing the complexities of enterprise transformation to guide the selection and creative application of appropriate tools and methods. The presentation flows from specific methodologies (source modules) to analyze dynamic effects, through the unifying time-dependent state-space descriptors of the enterprise, to the concept of an enterprise model that reflects the factors critical to enterprise transformation and the reality that some of these factors can be controlled, some can be influenced, and some are uncontrollable (the controlled, influenced, and uncontrolled [C-I-U] formulation). Multidisciplinary methods are particularly relevant to address the desired scope of policy, organizational, economic, and technical factors that must be coordinated in successful transformations of enterprise operations.

Chapter 5: Managing Enterprise Transformation Using ESE/A

This chapter describes how enterprise dynamics methods and the enterprise systems engineering and architecting (ESE/A) guidelines can provide value by relating the modeling and simulation methods to state-of-the-art management tools such as EA, activity-based management (ABM), and enterprise resource planning (ERP) systems. This architecture-based approach for governance of enterprisewide operations and transformation encompasses the POET elements and facilitates the engagement of technical, financial, and business managers in the operational and transformation processes. Architecture templates are described to couple dynamic analytical methods with ABM techniques to coordinate planning and management across the multiple agencies and stakeholders involved in complex undertakings. In this governance environment, enterprise resource management (ERM) accounting systems can be used to capture an always-current "state-space" description of financial and tangible resources of the enterprise. The approach is described using transformation examples that involve the coordination of multiple agencies (or firms). The multiagency environment is an increasingly important challenge for the enterprise, public or private, that operates in complex national and global markets with extended supply chains or reach.

The management approach is supported by ESE/A guidelines that can be tailored to specific enterprise transformation challenges. The diverse tools and methods described earlier must be selected and applied in a way that is appropriate to the scale and complexity of the ESE/A challenge. The ESE/A enabling process described here provides the necessary guidelines, and is based on experience with transformation programs, examples of which are described in the Section II case study applications. The process is flexible and adaptive in addressing the challenges and uncertainties in managing the transformation of complex enterprises in their unique forms.

Section II: Enterprise Modeling Approaches and Applications

In Section II, we present a series of ESE/A case studies based on published papers and technical reports on modeling the complex dynamics in specific enterprises to address critical issues in the acquisition of new operating capabilities. The case studies represent enterprise transformation challenges at a scale where the government role is paramount but with significant private sector and public involvement. The case studies employ a variety of modeling and analytical methods as appropriate to the specific enterprise transformation challenge they are addressing. Examining this variety in the context of the ESE/A process provides valuable and practical guidance for management and problem solving.

The case study applications are organized in relation to the defined scale of the enterprises starting with single agencies and corporations, and progressing through complex mission operations to multiple agencies and complex challenges of national scope with significant impacts on the economy. These are summarized in the following chapter descriptions.

Chapter 6: Simulation of Enterprise Architecture for a Business Strategy

This chapter applies the coupling of enterprise architecting and enterprise dynamics to strategic planning at the corporate level. The results of alternative investments of discretionary funds in business development and R&D are analyzed. This case study represents a major step forward in the foundational discipline that supports ESE/A through the concept of dynamic architecting. The application described implements a dynamic model and simulation with the comprehensive structure of an EA as defined by federal and commercial EA guidelines. This approach overcomes a major deficiency in the static nature of most architectures by joining this widely practiced enterprise planning and management method with enterprise dynamics.

Chapter 7: Reasoning on Technology Uncertainties for Enterprise Transformation

This chapter describes a method for reasoning about the likelihood of attaining a specified desired performance resulting from emerging dynamics and uncertainties within an enterprise or its environment. A basic principle is that mission performance occurs due to individuals understanding the dynamic interactions of people, process, and technologies. Individuals reason about dynamic interactions producing insights on likely performance attainment given the interactions. Individuals increase their understanding and their subsequent decisions and actions are changed because of their improved understanding.

The method uses evidential reasoning and multientity Bayesian networks (MEBN) to compose arguments about the relationship of technology initiatives to strategic outcome attainment. A federal program example illustrates the method.

Chapter 8: Optimal Control and Differential Game Modeling of a Systems Engineering Process for Transformation

This case study describes a unifying analytical framework for modeling enterprise dynamics in the ESE process across a range of enterprise and program types, and is demonstrated on a specific transformation program. The framework is a control-theoretic formulation that uses state and control parameters relating to the enterprise, stakeholders, and the environment with which the enterprise interacts. The framework expresses the individual self-interests of stakeholders and environmental players as a differential (dynamic) game, which is explored and interpreted in terms of game-theoretic Nash equilibrium solutions (Nash 1950). It provides a mathematical basis for relating or integrating the diverse ESE modeling methods employed in the application examples that follow.

Chapter 9: Hybrid Systems Dynamic, Petri Net, and Agent-Based Modeling of the Air and Space Operations Center

This chapter describes an innovative enterprise systems engineering effort to model the policy, organizational, and technical aspects of a mission-critical national defense operation. The application uses hybrid systems dynamics, Petri net, and agent-based multiscale modeling to understand the effect of operator–environment interaction and the global environment on Air and Space Operations Center (AOC) processes. The AOC process model (e.g., critical event process time and probability of errors) is linked to a global environment model that is driven by the political landscape in which the AOC operates.

Chapter 10: Nuclear Waste Management Strategic Framework for a Large-Scale Government Program

This case study presents an enterprisewide framework using the hybrid approach of a process-based materials flow model coupled with a systems dynamics influence diagram, or causal loop diagram. The objective of the work is to provide system-level insight into the U.S. Department of Energy's (DoE) responsibility for environmental cleanup of legacy nuclear waste. The focus is on the Savannah River site and all activities carried out in this enterprise from the receipt of nuclear materials through their processing to the shipment of the materials in forms suitable for safe long-term storage. The framework is used for exploring policy options, analyzing plans, addressing management challenges, and developing mitigation strategies for the DoE Office of Environmental Management (EM). The sociotechnical complexity of EM's mission compels the use of a qualitative approach to analysis to complement a more quantitative discrete event modeling effort. We use this analysis to drive scenarios for the model, pinpoint pressure and leverage points and develop a shared conceptual understanding of the problem space among stakeholders. This approach affords the opportunity to discuss dynamic phenomena in enterprise operations over a 25-year time horizon using a unified conceptual perspective, and is also general enough that it applies to a broad range of capital investment/production operations problems.

Chapter 11: International Trade and Commerce: Enterprise Systems Engineering and Architecting in a Multiagency Environment

The objective of this case study is to formulate and demonstrate a comprehensive planning framework for ESE/A in a complex international enterprise that is essential to the global economy. The framework, an integrated enterprise systems engineering workbench, is designed for multinational and multiagency enterprises, public and private, engaged in international commerce.

Chapter 12: Energy and Materials Systems as an Enterprise Systems Engineering Application: Planning and Analysis for the Economy's Infrastructure

This case study provides an example of a large-scale ESE challenge to the private sector and government, crossing major materials and energy-related sectors of a nation's economy. Major transformational initiatives have been proposed to deal with resource depletion and environmental challenges. The energy and materials systems underlying the physical infrastructure of national economies are complex and are addressed here through combined technical and economic analytical methods.

Chapter 13: Modeling the Nation's Healthcare System as a Dynamic Enterprise

This case study provides an example of a government ESE challenge to transform healthcare to provide greater access at a sustainable cost with effective outcomes. Healthcare sectors account for 18% of the U.S. economy with roughly half of that being funded through government programs. This application spans the POET aspects of the challenge at multiple scales ranging from demographics through specific diagnostic and therapeutic services, the structure of healthcare sectors and interactions with other sectors of the economy, to the overall economy. A framework is presented that helps to integrate analytics in this data-rich, but information-poor environment.

Epilogue: Enterprise Systems Engineering and Architecting— Lessons Learned and the Road Ahead

The epilogue reinforces descriptive enterprise dynamics as an essential part of enterprise planning and analysis for the acquisition and implementation of transformational technologies and processes to improve mission performance. This dynamic perspective of the POET aspects of the enterprise is a central feature of the ESE/A process. This sourcebook is a snapshot and assessment of the current state of an emergent field of national import and a platform for further research and operational applications.

References

Hybertson, D.W. 2009. *Model-Oriented Systems Engineering Science: A Unifying Framework for Traditional and Complex Systems.* Boca Raton, FL: CRC Press.

Nash, J.F. 1950. Equilibrium points in n-person games. *Proceedings of the National Academy of Sciences*, 36(1): 48–49.

Rebovich, G. and White, B.E. 2010. *Enterprise Systems Engineering: Advances in Theory and Practice.* Boca Raton, FL: CRC Press.

Stevens, R. 2010. *Engineering Mega-Systems: The Challenge of Systems Engineering in the Information Age.* Boca Raton, FL: CRC Press.

Editors

Kenneth C. Hoffman is a senior principal systems engineer at The MITRE Corporation's Center for Connected Government. He earned a PhD in systems engineering from the Polytechnic Institute of Brooklyn and a BS in mechanical engineering from New York University. He is engaged in the planning and analysis of energy, healthcare, and financial systems.

Ken has over 40 years of experience in R&D and executive management in manufacturing and service organizations. He was chairman of the Department of Energy and Environment, and director of the National Center for Analysis of Energy Systems, at Brookhaven National Laboratory. He led the development of energy system–economic models and their application to national energy R&D planning. In addition he was engaged in R&D on energy and materials technology and was project engineer on experimental facilities including the Brookhaven Solar Neutrino Observatory.

His career path from Brookhaven included executive management positions at the Mathtech Division of Mathematica Corporation (1980–1985) and (upon acquisition) at Martin Marietta Corporation (1985–1990), then to his current position at The MITRE Corporation, a not-for-profit organization that operates six federally funded research and development centers (FFRDCs).

Christopher G. Glazner is a lead information systems engineer at The MITRE Corporation's Center for Connected Government. He holds a PhD in engineering systems and a masters in technology policy from the Massachusetts Institute of Technology, and degrees in electrical engineering and Plan II from the University of Texas at Austin. He currently leads a group engaged in modeling and simulating government enterprises, capturing the interaction of process, policy, and organizational dynamics in an effort to build more effective government enterprises. He has worked with the Department of Veterans Affairs, Department of Energy, Internal Revenue Service, U.S. Courts, and the Department of Homeland Defense.

William J. Bunting is a principal information systems engineer at The MITRE Corporation's Center for Connected Government and earned a PhD in information technology from George Mason University and an MBA from the Wharton

School of the University of Pennsylvania. He supports enterprise transformations in federal agencies involved in border management, immigration, and defense.

Bill has over 30 years of experience in enterprise architecture, requirements engineering, and systems engineering. His focus has been on aligning technology investments to strategic goal attainment, effective use of enterprise architecture information, making valid enterprise business decisions, development of quality requirements, and the establishment of processes for enterprise architecture, requirements engineering, and organizational business decisions. He has supported a wide range of transformations such as executive-level decision making for army logistics, aligning technology implementation for entry screening of all cargo into the United States, and accurate requirements engineering for biometric-based identity management.

Bill has worked for several companies starting with the Boeing Company, Arthur Anderson, and Federal Data Corporations, and in 2002 joined The MITRE Corporation.

Leonard A. Wojcik is a project team manager at The MITRE Corporation's Center for Advanced Aviation System Development. He earned a PhD in engineering and public policy from Carnegie-Mellon University, an MS in physics from Cornell University, and a BA in physics and mathematics from Northwestern University. He is engaged in analysis of the Next Generation Air Transportation System (NextGen).

Len has over 30 years of experience in engineering and policy research and analysis. He analyzed air transportation safety and infrastructure construction R&D at the Office of Technology Assessment (OTA) of the U.S. Congress (1987–1988). He was director of research at the nonprofit, industry-supported Flight Safety Foundation (1988–1990). At MITRE (1979–1987 and 1990–present), he served as liaison to the Santa Fe Institute, and his engineering research and analysis span various large-scale military and civil aviation systems. Len's research includes applications of agent-based and economywide modeling to assess large-scale engineered systems.

Anne Cady is a director and chief systems engineer at MITRE's Center for Connected Government. She has BS degree majors in mathematics, chemistry, and biology and an MS in biochemistry. She has over 30 years of experience in technical leadership and management and hands-on development in all aspects of enterprise, system, and software engineering. Anne also has experience in modeling and simulation, database design and development, and system implementation.

Anne has led numerous design, analysis, development, and technical assessment projects for large complex systems in the U.S. Department of Defense (DoD), the intelligence community, Department of Homeland Security (DHS), and other federal government civilian agencies. She has led the implementation of enterprise resource planning systems in multiple organizations. She was principal investigator for a MITRE internal research and development project on emergency preparedness and response that focused on creating a dynamic incident management enterprise. Her current work focuses on the implementation and transformation of complex adaptive systems and enterprise ecosystems.

Contributors

Lindsley G. Boiney
MITRE
Bedford, Massachusetts

William J. Bunting
MITRE
McLean, Virginia

Anne Cady
MITRE
McLean, Virginia

Daniel B. Chamberlain
MITRE
McLean, Virginia

Fran Dougherty
MITRE
McLean, Virginia

Christopher G. Glazner
MITRE
McLean, Virginia

Kenneth C. Hoffman
MITRE
McLean, Virginia

Richard Hubbard
MITRE
Bedford, Massachusetts

Honora R. Huntington
MITRE
Chicago, Illinois

John James
MITRE
Bedford, Massachusetts

Joseph K. Jun
MITRE
McLean, Virginia

Dave Klein
MITRE
McLean, Virginia

Kristin Lee
MITRE
McLean, Virginia

Gregory A. Love
MITRE
McLean, Virginia

Patrick B. Mahoney
MITRE
McLean, Virginia

Paula Mahoney
MITRE
Bedford, Massachusetts

Jennifer Mathieu
MITRE
Bedford, Massachusetts

David H. Reid
MITRE
McLean, Virginia

Bradley C. Schoener
MITRE
McLean, Virginia

Samuel G. Steckley
MITRE
McLean, Virginia

Teresa A. Tyborowski
Department of Energy
Washington, DC

Mark Walters
MITRE
McLean, Virginia

Elaine S. Ward
MITRE
McLean, Virginia

Brian E. White
CAU-SES
Sudbury, Massachusetts

Leonard A. Wojcik
MITRE
McLean, Virginia

FOUNDATIONS AND CONCEPTUAL FRAMEWORKS

This sourcebook draws upon and describes a number of methods to provide a proven and structured approach, a framework, for Enterprise Systems Engineering and Architecting (ESE/A). It is directed toward analysts, managers, and decision-makers engaged in significant transformation of enterprises to operate at high performance levels in the delivery of products and services. It is also intended for the academic community for research and education purposes.

Representing, analyzing, and evaluating dynamic behaviors is a very important part of that framework. Enterprise Dynamics is a discipline for analysis of the complex interaction of policy, organizational, economic, and technical factors within an enterprise that affect the operations of the enterprise, and the acquisition and implementation of new capabilities for mission or business performance. The descriptive capabilities of Enterprise Dynamics address this need. They build on the concept of descriptive geometries and involve the mapping of the highly multidimensional enterprise state-space into the decision space for management of system and technology acquisition, policy formulation, and business operations. Several conceptual models are described to apply the discipline of Enterprise Dynamics.

The objective of Enterprise Dynamics is to develop and apply dynamic theory to acquiring and implementing technology, systems, and services for new operating capabilities in very complex organizational and operational environments involving Enterprise Systems Engineering (ESE) applications. Understanding the characteristics of the enterprise landscape and the need to support specific management decisions will increase the opportunities for success in this failure-prone endeavor.

Working Definition of an Enterprise. An enterprise is a purposeful social, technical, and economic undertaking designed to create value for its stakeholders involving:

- Policies, organizations (people), economics, and technology interacting with each other and their environment in operational processes as a complex system-of-systems to achieve goals
- An organization (single or multiagent, and possibly virtual) created for the undertaking
- A readiness to embark on bold new ventures

The dynamics of an enterprise are highly dependent on the mission and/or business characteristics (the enterprise landscape) of—and the perspective on—the enterprise required to address management issues and inform decision-making (operational, scalar, and temporal perspectives).

Working Definition of Enterprise Transformation. The transformation of an enterprise to a desired future state is a comprehensive managed change process driven by public (consumer) demands, competitive factors, economic forces, and technological opportunities. The primary transformation objective is significant performance improvements and the effective coordination of policies, organizations, economics, and technology (POET) in operational strategies and processes. Performance improvements include better products and services, cost, and quality. The dynamic and iterative change process and key elements addressed in this sourcebook as the discipline of enterprise dynamics are portrayed in Figure I.1.

Enterprise Dynamics Defined. Enterprise dynamics is defined in this sourcebook as a discipline that deals with the time-dependent interactions of the internal elements of an enterprise (policies, organization and staff, economic resources, and technology assets) and the characterization of its transformation to increase performance levels in a changing external social and market environment. The ability of the enterprise to internalize and influence aspects of the "external" environment, or to respond to uncontrollable elements, introduces further complexities dealt with in the foundations of enterprise dynamics outlined in Section I. Concepts and analytical methods for enterprise dynamics are drawn from engineering, economics, management sciences, and operations research to address complexities over the range of enterprise characteristics encompassed in the Section II case studies. The discipline of enterprise dynamics is a major contribution to the practice of enterprise systems engineering and architecting (ESE/A).

The ESE/A process is designed to enhance the timeliness and dynamic content of state-of-the-art management systems such as Activity-Based Management (ABM) and Enterprise Resource Planning (ERP) to enable the agile sense-and-respond enterprise.

Each of the case studies in Section II represents unique transformation challenges and draws upon applicable models and approaches from those described in

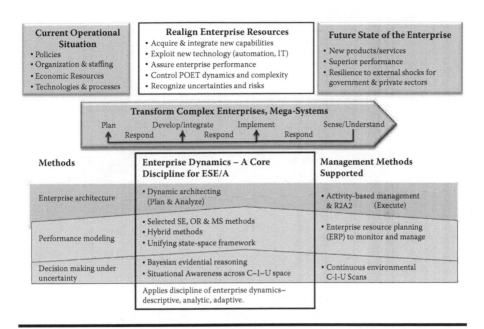

Figure I.1 Enterprise dynamics as a discipline supporting transformation.

Section I. The case studies represent "purposeful undertakings" of an enterprise. They are organized in relation to the defined scale of the enterprises starting with single agencies and corporations, and progressing through complex mission operations to multiple agencies and complex challenges of national scope with significant impacts on the economy.

Chapter 1

Defining Enterprise and Transformation Challenges

Kenneth C. Hoffman, William J. Bunting, and Anne Cady

Contents

1.1 Introduction

Government agencies and commercial entities are constantly evolving and transforming to respond to public demands, exploit opportunities, meet new market challenges, and adapt to a changing world. The networking of global supply chains for physical goods and information systems for financial services and operations management is based on new system technologies and business models that drive a high level of performance and operational excellence. Planning, acquiring, and implementing these integrated systems in organized processes require the application of systems engineering methods at the enterprise level: enterprise systems engineering (ESE).

5

An enterprise engaged in a dynamic mission involving multiple internal and external stakeholders operates as a complex adaptive system with policy, organizational, economic, and technology dimensions. Management decisions and the actions of the multiple stakeholders, as well as foreseen and unforeseen events, influence an enterprise's dynamic evolution. Systems engineering applied in this context spans the entire life cycle from research and development through the development and acquisition of new capabilities, transition to operations, operational use and maintenance, and finally to system retirement and replacement. This evolutionary life cycle is extraordinarily complex as uncertainties dominate along with unanticipated and uncontrollable events that influence the life cycle and the dynamic evolution of the enterprise. Enterprise systems engineering and architecting (ESE/A) addresses the complexities of this endeavor.

In this sourcebook, we describe and apply frameworks and analytical models of the enterprise as a complex system to improve the understanding of its dynamic elements and their interactions. Dynamics is a very important part of all engineering disciplines. Concepts of enterprise dynamics play a central role in this new discipline of ESE. The dynamic interacting elements of the enterprise can be broadly classified as the policies, organizations, economics, and technology factors in an operational process environment; we call these the POET conceptual elements. Specific enterprise transformation efforts must involve all of these elements with a balance appropriate to the specific enterprise. Planning, implementing, and managing the transformed enterprise require multiple skills and methods. This sourcebook focuses on the multidisciplinary methods drawn from management science, economics, engineering, operations research, and the social sciences that comprise a discipline of ESE/A.

1.2 Scope and Principles

The foundational principles underlying ESE/A are:

1. The modern enterprise engages in coordinated multiagency activities of public and private sector entities operating on a global scale with diverse activities, markets, and sources to deliver products and services.
2. Enterprise transformation is an ongoing process in sustainable organizations operating in an environment of constant and discontinuous change with associated risks as well as opportunities.
3. The planning and management of transformation require multiple skills and disciplines across policy, organizational, economic, and technology dimensions that can be drawn together and applied using a systems approach based on systems engineering, enterprise engineering, enterprise architecture, and management science.

4. Qualitative and quantitative information and methods are important to operational decision making and may be integrated and applied to the complex challenges of enterprise transformation using unifying techniques.
5. Emergent behaviors are an ever-present challenge to enterprise transformation and sustainability, and can be managed or exploited using theories and methods drawn from theory and experience with complex adaptive systems.

The enterprise as a purposeful socio-techno-economic undertaking involves multiple organizations and agencies activated by the coordination of policies, organizations (people), economics (budgets), and technology to provide services or deliver products to the government or public. This functional definition of the "enterprise" differs from the institutional definition of an enterprise as an agency or commercial entity. They are related in that an institutional enterprise is organized to carry out one or more purposeful activities.

The application of ESE/A in the Section II case studies focuses on the acquisition of new capabilities for enterprise transformation and operational excellence in both the targeted state and during transformation. The methods can be applied using the ESE/A process to guide the planning and managing of an enterprise as it implements new capabilities for the delivery of products and services in private and public sector endeavors. The results also provide quantitative benchmarks against which all kinds of enterprise programs, encompassing acquisition of capabilities and their operational use to carry out the mission, can be compared. Such modeling could also be used to support decision making and the selection of acquisition and management approaches for these extremely complex efforts.

Systems engineering methods have evolved as a formal discipline over the past 60 years to address emerging technical and socioeconomic challenges, often with new analytical techniques arising both from those challenges and from active research programs in computational methods. This evolution has extended the practice of systems engineering from purely technical systems as in automated control systems, to more complex domains such as energy systems, air traffic control, and combat systems. Major events and developments in the systems engineering timeline are outlined in Chapter 2 as a consistent trend toward ever-increasing complexity based both on the needs of modern society and on burgeoning analytical capabilities.

Engineering is defined as "the application of science and mathematics by which the properties of matter and the sources of energy in nature are made useful to people" (Merriam-Webster). As an engineering discipline, systems engineering has consistently employed quantitative methods, recognizing, of course, that not all aspects of the system challenge can be quantified.

Early applications in the 1950s and 1960s focused on applying physical sciences to transportation, nuclear power systems, weapons systems, and the space program. Another major aspect of these applications involved the integration of multiple engineering disciplines: chemical, mechanical, civil, electrical, and nuclear. The success of applications in these predominantly technical areas created an appetite for the application of systems engineering to social challenges involving institutions, enterprises, and the public. This expanded scope into "social engineering" was not nearly as successful and offers significant lessons. The primary lesson applied in this sourcebook to the multidisciplinary field of enterprise systems is to return to basic engineering principles, but in a broader multidisciplinary context that builds on the applicable sciences, physical and social. Enterprise systems are addressed in management science, organizational theory, process engineering, economics, operations research, and other disciplines. The systems engineering approach provides a proven method of integrating the multiple perspectives and insights arising from these disciplines and applying them to the challenges faced in transforming and modernizing public and private enterprises to address markets and public interests.

This sourcebook describes a state-of-the-art approach to ESE/A in two parts:

I. Foundational and Conceptual Frameworks
II. Enterprise Modeling Approaches and Applications

ESE/A process guidelines are formulated with sufficient generality to apply to a wide range of enterprise challenges, government and commercial. The process guidelines address many types of programs involving dynamic interactions among multiple stakeholders in the enterprise around policies, organization, economics, and technology.

1.3 Conceptual Frameworks

A conceptual framework uses landscapes and architectures to characterize the specific nature of an enterprise. Landscapes characterize the form and objectives of the enterprise—in terms of efficiency, flexibility, and adaptability—as well as the nature of the operational processes to deliver repetitive services, specialized services, or adversarial actions. Enterprise architectures (EA) capture the key activities in the enterprise, the supporting services (logistic and informational), and the network applications and computing resources needed.

These artifacts help to define the scope and structure of the enterprise to be addressed as a complex system. Systems theory requires a precise definition and description of system boundaries and relationships within and across the boundaries, wherever drawn.

No characterization of an enterprise is complete without an explicit representation of the decision processes that guide its transformation. An enterprise has a set of internal and external factors that are under its direct control, factors that it can influence, and factors that are uncontrolled, again both internal and in the external environment within which the enterprise operates.

Management decisions (and equally important "nondecision" decisions) rely on the enterprise sense-and-respond capabilities where the situation is monitored and measured across the controlled, influenced, and uncontrolled (C-I-U) space. The C-I-U paradigm is captured in the ESE/A process explicitly in a control-theoretic framework, specifically the C-I-U theoretic model.

Enterprise resource planning (ERP) methods provide a representation of the organization, activities performed, and resources utilized in the enterprise. Activity-based management (Player and Keys 1999) is often coupled with ERP to provide portions of a sense-and-respond capability. These methods along with EAs play an important role in ESE/A.

Geographic information systems (GIS) are used to represent the physical location of enterprise assets, markets, or operational environments and the movement of products, services, and resources. They play an important representational role in the ESE/A process as is demonstrated in several of the case studies.

The ESE/A process guidelines outlined in Section I include the following principal elements:

- Characterization of the complex enterprise and representation of the processes, objectives, and important internal and external relationships
- Multidisciplinary analysis of the cost and performance parameters across policies, organizations, economics, and technology (POET) elements for alternative transformation strategies and solutions using quantitative and qualitative methods
- Analysis of stakeholder interests and influences, and organizational roles and responsibilities
- Implementation of acquisition or transformation strategies and systems
- Ongoing monitoring and management to achieve objective levels of improvement in operational performance

1.4 Enterprise Dynamics Processes

Enterprise dynamics is a discipline for analysis of the complex interaction of policy, organizational, economic, and technical factors within an enterprise that affect the operations of the enterprise and the acquisition and implementation of new capabilities for mission or business performance. These include:

- *Performance Dynamics* in acquisition programs and in ad hoc proactive and reactive missions and business initiatives
- *Organizational and Behavioral Dynamics* involving collaboration across agencies and among stakeholders to perform complex acquisitions and missions

■ *Information Dynamics* to collect and exchange information that supports sense-and-respond decision processes during acquisition and the transition to operations

■ *Technology Dynamics* governing the maturity, rate, and degree of difficulty in employing advanced technologies for improved enterprise performance

A comprehensive study of complexity in the forward-looking transformation to network-centric operations enabled by new information systems, sensors, and other technologies was performed by the National Research Council (NRC 2006). This study focused on U.S. Army applications but is referenced as "... an exercise in coping with complexity."

Although the focus was on network-centric applications in the U.S. Army, the network-centric paradigm is being applied to transform government services, commercial supply chains, and healthcare and the conclusions and recommendations apply directly to that wide range of enterprise domains. A major point of interest in the NRC report is the statement that "... trying to implement net-centric operations capabilities as envisioned ... is like trying to design and build a modern combat jet aircraft without resorting to the science of fluid dynamics." This statement reinforced the importance of the enterprise dynamics research program that was in progress at the time and suggested a strong focus on transformation challenges in network-centric or network-enabled enterprises.

The NRC report identified a number of challenges in this class of ESE programs, which motivated the ESE/A process described in the following chapters, including the following:

■ Lack of overall integrating architectures and systems engineering for enterprise networks

■ Inadequately trained, educated, and certified personnel and network users

■ Network management and lack of joint network configuration management

■ Network security and information assurance

■ Requirements to model, simulate, and test large networks before deployment

■ Fusion of multiple sensors and sensor types across the network for renewal time decision making

■ Design and integration of individual service networks

■ Understanding of the relationship between network structure and complexity and its impact on organizational design and individual and unit behaviors

■ Energy-efficient electronics to reduce soldier loads and simplify logistics support

The landscape, modeling, and architecting element of the ESE/A process must address these challenges, and must encompass policy, organizational (people), economic, and technology elements of enterprise dynamics in planning and managing the transformation to greatly improved operational levels.

1.5 Modeling Approaches and Applications

A very wide range of disciplines and models can be drawn upon for ESE/A analysis of POET dynamics. Specific models and methods applied to enterprise system engineering depend on the system description, the objectives of the ESE exercise, and the characteristics of a specific enterprise. The characterization of the enterprise and its transformation challenges uses the conceptual framework discussed in Section 1.3 and serves to identify the applicable disciplines and the specific modeling capabilities required to deal with factors that can be quantified as well as critical qualitative factors.

Several example applications are presented covering various scales and other characteristics of an enterprise ranging from a specific operation to a major sector of the U.S. economy, as well as various operational environments, including the delivery of repetitive services or products and situation-dependent operations of a highly variable nature. The examples encompass many analytical methods that may be drawn upon to address complex enterprise system challenges in their appropriate roles.

The emphasis in ESE/A modeling is to capture quantitative and qualitative factors that affect enterprise performance and the ability to achieve operational improvements through transformation programs. Predictive capability is greatly desired in such models, however, any such capability is highly conditioned on the assumptions and factors considered. Applications of models in the ESE/A process emphasize descriptive and normative capabilities to analyze alternative decisions and actions and to inform managers and other stakeholders.

The variety of modeling methods and disciplines applied in the case study examples of Section II includes control theory, system dynamics, agent-based models, Bayesian probability theory, highly optimized tolerance, Fourier series, and game theory.

No treatise on ESE can be complete without addressing the social aspects of planning, modeling, and analysis. All stakeholders must be drawn into the process to reveal objectives and priorities, and to communicate the results of the ESE process and analytical work. The closing chapter in Section I provides an approach to governance and ESE/A process guidelines that include engaging stakeholders in the effort.

1.6 Conclusion

The definition of the enterprise as a purposeful socio-techno-economic undertaking involving multiple organizations and agencies can be applied at multiple scales.

In this era of constant and dynamic change, enterprises must undergo continuous transformation to exploit opportunities and respond to markets, a complex system problem. Such transformation requires the coordination of dynamic processes involving policies, organizations, economics, and technology to provide services or deliver products to the government or public; this is the theme of enterprise dynamics that is the major topic of this sourcebook.

Before delving into the methods and models of enterprise dynamics it is instructive first to review the evolution of ESE/A as applied to complex system challenges.

References

Merriam-Webster Online Dictionary, http://www.merriam-webster.com/dictionary/enterprise.

NRC (National Research Council), Committee for Future Army Applications. 2006. *Network Science*. Washington, DC: The National Academies Press.

Player, S. and Keys, D.E. 1999. *Activity-Based Management*. 2nd ed. Hoboken, NJ: John Wiley & Sons.

Chapter 2

From Systems Engineering to Enterprise Systems Engineering

Kenneth C. Hoffman

Contents

2.1 Introduction—Looking Back and Looking Forward

Looking backward, systems and the enterprises they supported were viewed as relatively independent entities and were treated that way. First, industrial engineers and then systems engineers have been quite successful in dealing with them as well-bounded systems with repetitive and predictable operational characteristics. These system characteristics are changing in significant and fundamental ways. Commercial enterprises addressing important markets and government enterprises addressing critical national problems increasingly operate in a collaborative heavily networked environment, networked in a technical, organizational, and operational sense. These characteristics are descriptive of a complex adaptive system, a system of systems, and an enterprise system; all of these terms apply.

Looking forward, there is a growing trend toward the concept of net-centricity and connectivity in a global economy as indicated by a study conducted for the army by the National Academy of Engineering (NAE) on "network science" (NRC 2005). The study conclusions reinforce the focus of this sourcebook on the broad field of enterprise dynamics. Their multilevel concept of enterprise networks encompasses social, cognitive, information, and physical levels and encompasses the policies, organizations, economics, and technologies (POET) elements of the enterprise used here. This sourcebook also highlights enterprise architectures (EA) as a primary information repository for enterprise landscape perspectives. Selected contents of such architectures can be mapped to the NAE levels as follows:

2.1.1 Social

- Socioeconomic environment that the enterprise serves
- Governance of the enterprise involving external and internal stakeholders
- Management and organizational interactions within the enterprise and with related agents/agencies

2.1.2 Cognitive

- Business performance
- Workforce behaviors
- Business rules and processes

2.1.3 Information

- Sensing of the external environment
- Sensing of internal status
- Business information resources

2.1.4 Physical

- Facilities
- Technical infrastructure

A variety of additional tools for planning and analysis is needed for the proper use of the rich but static information contained in EAs. This more extensive activity is referred to as enterprise system engineering and architecting (ESE/A). Following is an historical perspective on highlights in the long-term evolution of the field of systems engineering in response to challenges.

2.2 Foundations of Systems Engineering

Concepts of general systems theory, or system science, were utilized in the origin and continued evolution of systems engineering. Although useful, these concepts were often highly abstract and needed to be incorporated in a systems engineering discipline to support practical applications. An important and distinct foundation of systems engineering up to the 1990s is found in control theory (Kucera 2000):

> Control theory and engineering have witnessed dramatic achievements throughout this century. Recall the stability theory of Lyapunov from the beginning of the century, the conception of three-term or PID (Proportional-Integral-Derivative) controllers in the 1910s, electronic and pneumatic feedback amplifiers in the 1920s, Nyquist and Bode charts of the 1930s, and Wiener's cybernetics of the 1940s. Then came the 1950s and Bellman's principle of optimality, Kalman filter of the 1960s, adaptive control in the 1970s, robust control in the 1980s, and the hybrid control systems of the current decade.

The control-theoretic methods were particularly appropriate for applications to predominantly technical systems such as automated manufacturing, weapons systems, and nuclear power.

Academic origins of systems engineering are found in industrial engineering, which drew upon other engineering disciplines for industrial and civil applications. The incorporation of engineering economics in the curriculum was especially noteworthy, an addition that was often ignored, but proved highly beneficial to all engineering disciplines.

The applications of industrial engineering to such areas as public works and manufacturing clearly drew on system principles and are fundamental efforts for the discipline of systems engineering. In a period of constrained resources there was a reasonable balance between the economic (cost) and technological elements of those programs. Manufacturing firms and local governments had restrictive budgets and could not print money, so costs were dealt with very seriously as a constraint. This may be contrasted with budget "discipline" at the federal level.

When operational aspects of new technologies such as radar became a focus of attention in World War II, it was clear that adversarial strategies and human reactions were of great importance and the field of operations research (OR) developed rapidly to address human, organizational, and operational aspects of systems. Systems engineering drew on OR methods, and vice versa. They shared many common elements in their respective timelines as evident in an OR timeline (Gass 2002) that is integrated in the following section with the systems engineering timeline to illustrate the significant cross-fertilization that took place. Specialized areas related to OR such as management science and organizational science emerged and are among the more important disciplines that need to be drawn upon in a multidisciplinary ESE/A process. Certainly during WWII, and in defense programs thereafter, advanced capabilities to deal with threats to the nation took precedence over the economic elements.

OR, post WWII, is sometimes characterized as emphasizing theory and methodologies, whereas systems engineering emphasized applications in keeping with engineering traditions. Systems engineering is generally offered as an interdisciplinary program within schools of engineering. Curricula have evolved from basically a control theory core through the incorporation of OR methods to a predominantly software development curriculum in response to program application challenges.

2.3 Evolution of Systems Engineering—Timeline

To the extent that systems engineering-like methods were applied in the early stages of this timeline, they generally emphasized the purely technical aspects of the problem separately from policy and organizational aspects of the enterprise. For example, the timeline for relevant systems engineering-related programs from the late 1930s to the 1990s is summarized as follows.

> 1938—NYC-LI Highway and Park System (Caro 1975). Often cited to illustrate emergent behavior in complex adaptive systems (e.g., unforeseen outcomes of suburbanization). Some native New Yorkers who enjoyed the highways and parks have a much more positive view of this era and the role of Robert Moses than that expressed by Caro.
>
> This period is exemplified by the technological vision of the Futurama exhibit at the 1939 NY World's Fair (Figure 2.1).
>
> 1940s—Bell Labs. First reference to systems engineering; see History (INCOSE 2010).

Figure 2.1 Technological vision of a highway system of the future, Futurama Exhibit, NY World's Fair, 1939. (Reprinted from nts.btl.gov, Public Domain.)

1946—RAND Corporation. Founded by the U.S. Air Force and created systems analysis, an important part of systems engineering (INCOSE).

1953—University of Pennsylvania. First graduate systems engineering curriculum.

1950s—Maturing of optimal control theory (related to Dantzig and Gass optimization work).

1954—Nautilus submarine. Launched. Origin of nuclear navy: an early ESE/A example with the parallel development of submarines, nuclear propulsion, and the supporting nuclear fuel cycle.

1956—CPM/PERT (Critical path method/program evaluation review technique). Developed as a systems engineering planning and management method (based largely on Nautilus experience).

1957—International Federation of Automatic Control (IFAC). Founded.

1958—MITRE. Established for systems engineering management of semiautomatic ground environment (SAGE) the first major real-time computer-based command and control system for air defense.

1961—Air Force Systems Command (AFSC). Established in recognition of unique system acquisition and management challenges, and absorbed the Air Research and Development Command (ARDC).

1961—Systems dynamics, feedback control theory. Applied to industrial dynamics (later, urban dynamics in 1966, world dynamics in 1972).

1963—SAGE. Fully deployed.

1969—Apollo 11 mission.

1969—ARPANet. First nodes operable.

1970s—Social engineering, limits to growth, world dynamics. Applied system dynamics with a wide scale. The method was overextended in the opinion of some.

1970s—Commercial large-scale integrators. Positioned for large complex systems contracts (formalized Apollo acquisition experience).

1971—DoD 5000 Series. Initiated for acquisition management (persistent evolution up to today).

1972—International Institute for Applied Systems Analysis (IIASA). Founded in Laxenburg, Austria to address complex multinational system challenges. Secretariat of IFAC colocated IIASA in 1975.

1980s—Software Engineering Institute. Published capability maturity models (CMM).

1987—Zachman Framework for Information Systems Architecture. Comprehensive characterization of the information systems operating across the enterprise, subsequently evolved to EA. See applications by Hoffman and Melancon (1988).

1990—International Council on Systems Engineering (INCOSE). Established.

1992—AFSC (Andrews AFB) and Air Force Logistics Command. Merged into Air Force Materiel Command (Wright-Patterson AFB).

Defense programs played a major role in the evolution of systems engineering over this timeframe with an emphasis on weapons system acquisition and ever-increasing attention to software engineering efforts. The acquisitions were characterized as major research and development efforts culminating in production systems, usually for a single customer (a monopsony). The emphasis on capabilities to protect the nation had a clear effect on the evolution of systems engineering and led in too many cases to imbalances in cost management that must be addressed in ESE/A.

2.4 Emergence of an Enterprise Systems Perspective—1990s to Present

To address significant cost, schedule, and performance shortfalls in many large complex systems and programs, it was recognized that systems engineering methods required significant augmentation to deal effectively with all aspects of a major program. Technical and nontechnical factors had to be dealt with in a more integrated way, drawing on a wider range of physical and social sciences with an enterprise perspective as a central theme. Increased attention was required regarding the objectives and perspectives of multiple stakeholders in the enterprise. Some highlights of this emergent period include:

1990—Initial applications of the Zachman Framework for Information System Architecture (1987) at the EA level across government agencies and the private sector.

1995—Enterprise Resource Planning (ERP) consisting of a bundled software application for the extended enterprise (builds on manufacturing resource planning (MRP) and supply chain concepts circa 1990).

2000—John Sterman's book on business dynamics extended earlier systems dynamics writings of Forrester and Meadows on system, urban, and global dynamics into the realm of enterprise dynamics.

2000s—Central Artery/Tunnel (CA/T) "Big Dig" experience in Boston revealed and reinforced continuing shortcomings in addressing extremely complex urban system challenges.

2005—Enterprise systems engineering (ESE) concepts emerged as research topics and potential solutions to complex programs (MITRE, MIT, University of Illinois). This sourcebook is an outcome of those research efforts on enterprise dynamics and programmatic applications.

2007—Lean advancement initiative (LAI) established at MIT to address manifold challenges of enterprise transformation.

The systems engineering timeline illustrates the constant enlargement of the scope and scale of systems engineering and the application of ever more powerful analytical methods appropriate to that larger scope and scale. Successes and failures are numerous and the ever-present risk of overextending or misapplying systems engineering methods must always be considered.

2.5 Future of Enterprise Systems Engineering, and Some Alphabet Soup—SoSE, CAS

Emerging challenges of globalization, security, the economy, and improved social services require significant transformations in private and public sector enterprises. Increased capabilities to address these very complex issues are enabled by new technologies that enhance the effectiveness of physical and information networks. The complexity of the challenges motivates the need for methods that draw upon the physical and social sciences and apply the robust capabilities of those disciplines: the integrating applied discipline of ESE. ESE must raise the capabilities of its methodology to deal with the complexities of the challenges.

Efforts to address the important, but more qualitative, aspects of systems such as stakeholder behaviors, organizational theory, complexity, and decision making require new methods. These aspects are of particular importance in enterprise systems, and are a major aspect of the POET scope of the ESE process that blends qualitative methods with the more traditional quantitative methods of an engineering discipline. Enterprise-level programs generally involve some mix of private sector, federal, and local government and, with increasing frequency, foreign partners. These multistakeholder programs require a well-balanced treatment of POET elements. Earlier systems engineering efforts reveal emphasis on one or more of those elements, usually for good reason, occasionally as an oversight. The ESE process must address all POET elements, with emphasis and priorities still dependent on the characterization of the specifics of the enterprise and the transformation program.

Previous efforts in systems engineering and its extensions to systems of systems engineering (SoSE) in an enterprise context were extremely challenging and many instances of such complex efforts have been problematic (Bar-Yam 2003; Hoffman et al. 2007). The systems engineering method must be tailored and extended with greater POET dynamic capabilities to deal with these complex adaptive systems (CAS). To cite a single, well-documented SoSE example dating back to the 1990s, the U.S. Federal Aviation Administration's (FAA) multibillion-dollar advanced automation system (AAS) program to modernize the U.S. air traffic control system took place in a context of major technical and management challenges together with unclear requirements in key areas, and was only partially completed, with major delays and cost overruns, when the program ceased to exist as originally conceived (Boppana et al. 2006). This reference presents system dynamics (SD) and highly optimized tolerance (HOT) models of SoSE processes applied to the AAS program, and poses the question of whether the complexity of SoSE in an enterprise context can be adequately represented and studied with such models. These models are described along with the other multidisciplinary enterprise dynamics methods.

2.6 Conclusion

This survey of the applications of systems engineering to larger and more complex challenges exhibits an ongoing evolution in the scope of transformation challenges and the willingness to embark on them. The evolution of modes and methods has been enabled by ever-increasing computing and analytic power, too often without commensurate success. This broad survey indicates that improved frameworks for planning and management are needed and can benefit from greater attention to dynamic models and methods employed in an ESE/A framework.

References

Bar-Yam, Y. 2003. When systems engineering fails—Toward complex systems engineering. In *International Conference on Systems, Man & Cybernetics*. Vol. 2: 2021–2028, Piscataway, NJ: IEEE Press.

Boppana, K. et al. 2006. Can models capture the complexity of the systems engineering process? Presented at the New England Complex Systems Institute (NECSI) International Conference on Complex Systems (ICCS2006), Boston (June 25–30).

Caro, R.A. 1975. *The Power Broker: Robert Moses and the Fall of New York*. New York: Vintage Books.

Gass, S. 2002. Great Moments in HistORy, *OR/MS Today*, 29(5). http://www.lionhrtpub.com/orms/orms-10-02/frhistorysb1.html.

Hoffman, K.C. and Melancon, J. 1988. An information systems architecture for manufacturing/distribution enterprises. In *Proceedings of the ASME Manufacturing International 88*, Atlanta.

Hoffman, K.C. et al. 2007. Descriptive enterprise dynamics—A multi-disciplinary unifying framework. Presented at the Fifth Annual Conference on Systems Engineering Research (CSER), International Council on Systems Engineering (INCOSE), Hoboken, NJ.

INCOSE (International Council on Systems Engineering). 2010. http://www.incose.org/practice/whatissystemseng.aspx.

Kucera, V. 2000. Automatic control: Past, present, and future. *ERCIM News* (online edition), 40. http://www.ercim.org/publication/Ercim_News/enw40/keynote40.html

NRC (National Research Council). 2005. *Network Science*. Washington, DC: National Academy of Sciences. http://darwin.nap.edu/books/0309100267/html/.

Chapter 3

Foundations of Enterprise Systems Engineering and Architecting

Christopher G. Glazner

Contents

3.1 Introduction

The current state of systems engineering (SE) and enterprise architecting (EA) is reviewed in this chapter along with enhancements to portray an integrated approach to enterprise systems engineering and architecting (ESE/A). The objectives of the enhancements described here are to advance the state of EA and SE practice from predominantly static and deterministic approaches to dynamic and probabilistic ESE/A processes.

This integrated approach draws on concepts of enterprise dynamics using models to describe mission performance, organizational change, cost elements, and technologies (Glazner 2009). Managing and enhancing the performance of public and private enterprise activities is the subject of architecting and modeling to apply the disciplines of management science, probability theory, organizational theory, systems engineering, and socioeconomics. Each discipline addresses specific dynamic elements of the enterprise from a unique perspective and, using hybrid modeling approaches, contributes a robust foundation for ESE/A that addresses the complexity and dynamics of the interacting policy, organizational, economic, and technical (POET) elements of the enterprise. This field has been fertile ground for research and applications over the past few years and is sufficiently mature to identify foundational concepts.

The chapter begins by reviewing the application of complexity theory to the enterprise, highlighting principles such as "systems thinking," near-decomposability, scale, and perspective and how they can be brought to bear in the analysis of enterprise behaviors. Next, the concept of frameworks is introduced as a holistic construct for managing enterprise complexity, with an eye toward its application for understanding enterprise dynamics. The chapter concludes with a discussion of how frameworks and the theories and constructs of organizational science can be brought to bear for an understanding of the complexities of enterprise dynamics.

3.2 Complexity in Enterprises

Organizations have long been considered "complex" in the casual use of the term, however, the study of them as *complex systems* did not begin until the 1950s and 1960s as a result of research done in the fields of general systems theory (GST) and cybernetics. These contemporary fields together gave rise to modern notions of systems and complexity, although they studied different aspects of complexity in systems. GST sought to identify and understand common structures, behaviors, and attributes of many different kinds of systems across scientific disciplines, ranging from biology to physics to sociology (von Bertalanffy 1956). Cybernetics studied the capability of systems for self-regulation through feedback and environmental sensing. Although cybernetics often focused on the study of technical systems, its followers applied their analyses equally to organizations and other sociotechnical systems.

The term *complex system* was used by both GST and cybernetics to describe a general class of system that exhibited behaviors that were difficult to predict. Herbert Simon (1962, p. 468) notes that "by a 'complex system' I mean one made up of a large number of parts that interact in a non-simple way. In such systems, the whole is more than the sum of the parts, not in an ultimate, metaphysical sense, but in the important pragmatic sense that, given the properties of the parts and the laws of their interaction, it is not a trivial matter to infer the properties of the whole."

Complex systems can be technical or social systems, or both; their key feature is the interaction of their parts. Even cybernetics, which tended to focus on technical systems such as fire control radars, saw how analyzing organizations in terms of the interaction of their major elements could help in understanding their behavior. Norbert Wiener, one of the fathers of the cybernetics movement, noted that "we must consider (organizations) as something in which there is an interdependence between the organized parts" (Wiener 1956). If the interdependence of the parts could be understood and characterized, the behavior of the system could be understood.

3.2.1 Applying Complexity Theory to Enterprises

The study of complex systems requires a balance of contrasting perspectives depending on the context of the system: holism versus reductionism, top-down versus bottom-up. The perspectives and analytical approaches taken in a study of a complex system are contingent on the system under study and the particular dynamics being investigated. There is no single approach that is best suited to the study of all complex systems and for all questions (Mingers and Gill 1997).

The study of enterprises as complex systems presents researchers and managers alike with a broad array of challenges. Enterprises may be designed from the top down or evolve from the bottom up and in all cases exhibit emergent behaviors requiring a response. They are both centrally directed and dependent on the actions of agents who make locally rational decisions dependent on their local incentives. Accordingly, different bodies of research and practice have been built around different approaches for managing the complexity of the enterprise.

3.3 Organizational Science

At its core, organizational science is the study of the structure, behavioral dynamics, and design of organizations. It seeks to identify scientific principles that can be employed to describe how organizations are structured and how they behave given their environments and offers suggestions on how new organizations can be designed. Organizational science uses the organization as its unit of analysis. This can be seen in contrast to the study of enterprises, which is often more practically oriented and uses the enterprise as its unit of analysis, extending the organizational construct to incorporate an extended value chain that includes stakeholders and its environment.

Although both the "rational" and the "natural" perspectives of organizational science have contributed greatly to the understanding of organizations, the "open systems" perspective is more suited to understanding the broader concept of enterprise. The open systems perspective places an emphasis on challenging the boundaries of the organization while also examining the nature of its interdependencies (Thompson 1967). It recognizes that organizations often exhibit a high degree of loose coupling (Cyert and March 1963), which can allow the system to be more adaptive and flexible (Orton and Weick 1990). This perspective is most likely to include analysis of an enterprise's environment, processes, technologies, and products in addition to organizational forms and decision-making (Scott and Davis 2006). The open systems perspective tends to take a more holistic approach to analysis than either of the other two perspectives of organizational science. From the vantage point of the enterprise architect, the primary areas of interest within the open systems perspective of organizational science are contingency theory and the closely related field of organizational design.

The field of organizational design capitalized on the work of contingency theorists with the aim of providing prescriptive guidance to those wishing to design organizations. In addition to the study of organizational contingencies, proponents of organizational design advocated the study of such areas as work flows, control systems, information processing, planning mechanisms, knowledge transfer, and their interactions. The field of organizational design tends to be more applied than most other fields within organizational science; organizational designers seek to change and improve organizations from a managerial perspective rather than describe and understand them from a theoretical perspective (Scott and Davis 2006).

An organizational designer embraces the complex systems philosophy of "the importance of treating the system as more than the sum of its component elements" (Scott and Davis 2006). As such, heavy emphasis was placed on understanding the fit and interdependency of an organization's design components. Thompson, a contingency theorist, is considered an early pioneer of organizational design for his work in identifying and categorizing interdependencies in organizations and the dimensions of coordination in organizations (Thompson 1967).

3.3.1 Computational Organizational Science

Quantitative analysis of the architecture of enterprises is largely missing from both the contingency theory and organizational design literatures. This is a critical connection to make in managing organizational change in coordination with other elements of a transformation program. There are few tools to quantitatively assess trade-offs made in organizational design, or to assess one architecture against another. Due to the inherent complexity of enterprises, closed-form mathematical or logical analysis of designs is limited to only the absolute simplest structures, such as the analysis of the complementarity of structure and strategic fit performed by Milgrom and Roberts (1995). Analysis of modest real-world organizational forms, processes, and behaviors requires the use of computer simulation models capable of modeling the enterprise as a network of interacting adaptive components over time.

Research organizations' interest in using simulation models is not a recent development. Going back several decades, organizational design theorists advocated the use of simulation models where "all the variables and relationships of interest are linked as understood in a model and then the manager-analyst-researcher manipulates certain ones and observes how others change as the simulation of the system plays out" (Swinth 1974). Despite the desire to use simulation models, it was many years before both simulation techniques and computational power matured enough to be applied to the types of problems of interest to organizational scientists.

Carley (1995) was among the first to advocate the creation of a new discipline within organizational science, to be known as "computational organizational science," arguing that virtual experiments employing simulation models could be used for the purpose of creating and testing organizational theory, especially for the application of complexity theory to the organization to understand how the models adapt and evolve. Simulation models allow organizational researchers to create controlled, easily manipulated experiments where real-world "natural experiments" inside organizations are not feasible due to cost, scope, or ethical considerations (Fowler 2003). Techniques such as agent-based modeling allow the creation of simulation models that capture the behavior of an organization at multiple scales by simulating the interaction of its elements. Such models allow analyses of how organizations search, adapt, and learn in changing environments.

Computational organizational science has come to focus primarily on taking a complex *adaptive* systems view of organizations and has used simulation models to establish theories for how organizations evolve and adapt in uncertain environments, especially from a bottom-up perspective (Anderson 1999). Although still in its infancy, computational organizational science shows much promise at developing theory that helps researchers better understand how enterprises adapt at multiple scales to their changing environments.

Computational organizational science primarily uses simulation models in a theory-generative or theory-testing role rather than as an analysis tool in the process of organizational design. The models employed to date have been highly abstract "toy" models that served to prove an hypothesized mechanism, rather than as a simulation of a real-world organization or system. Although such simple models have proved valuable in the creation of new organizational theories governing the evolution of organizational design (Rivkin and Siggelkow 2003; Ethiraj and Levinthal 2004), they cannot be considered tools that can be used by managers as analysis tools in organizational design. These "toy" models are far too abstract for managerial application. The aim of computational organizational design is to develop new theory, not to create new architecting and analysis tools.

Breaking from the tradition of using simulation modeling exclusively for theory generation, Levitt (2004) argues that the development of analysis tools for organizational design has been hampered by the lack of robust theory that supports it. He then argues that this is changing, and that two major strands of organization research have provided the foundations for a theory strong enough to support analysis tools that can be used to design organizations:

1. The information-processing view of the organization (March and Simon 1958; Galbraith 1973) provides the framework for agent-based simulation models at the micro levels through an understanding of information pathways and local incentives.
2. Contingency theory provides the macro-level propositions that relate structural form to context, based on empirical observation.

By bringing together theory of organizations from both top-down and bottom-up perspectives, Levitt argues that simulation models can link these theoretical perspectives and should be used to analyze the potential performance of organizational design in different environments. He states that simulation technologies have come of age, making the simulation of organizational design both technically possible and theoretically supportable.

3.3.2 Resolving Gaps in Organizational Theory for Effective Dynamic Organizational Design

Organizational science as a whole has strived to create rigorously developed, generalizable theory with broad explanatory powers. Organizational theory is capable of providing some normative guidance to managers seeking to direct the transformation of their enterprise, but it does not provide enterprise managers and architects with the tools needed to analyze and apply the theory to their own enterprises quantitatively. Ideally, managers would adapt and change their organization's design based upon a careful consideration of current organizational theory, however, studies have repeatedly shown that managers tend to adapt their organization's design using a costly "trial and error" process based upon prior experience rather than rely on theory (Tatum 1983). Levitt argues that without the ability for managers and architects to analyze potential changes to their own enterprise's architecture before they are implemented, it is impossible to create an effective architecting process. Without an effective process, the enterprise will revert to using the ad hoc trial and error approach (Levitt 2004).

Organizational science has been missing frameworks and theories that can tie together the richness and rigor of its research in a coordinated manner that makes the full body of knowledge more useful to the enterprise leaders managing the development of their enterprises. Until such time as a unifying "enterprise science" can mature and develop, enterprise leaders must look elsewhere to find useful guidance for stitching together stovepiped bodies of knowledge in order to create tools and models to help them better understand and manage enterprise behaviors. Enterprise architecting provides a practical framework for resolving the gaps in quantitative organizational science for application to enterprise transformation.

3.4 Enterprise Architecting

The practitioner-oriented field of enterprise architecting offers a holistic approach to managing an enterprise's complexity. Enterprise architecting holds that enterprises, as complex systems, can be "architected" in an organized fashion that will give better overall results than an ad hoc or piecewise approach to organization and design (Bernus 2003). To architect an enterprise is to create and document a specific abstraction of it that describes its fundamental mission, organization, and resources either in its current state or as a desired future state. This abstraction is known as the enterprise architecture.*

The ANSI/IEEE definition of EA states that it is "the fundamental organization of a system, embodied in its components, their relationship to each other and the environment, and the principles governing its design and evolution" (ANSI/IEEE 1471: 2000). A second, more complete definition is given by Bozdogan, who states that "enterprise architecture is an abstract representation of a 'real-life' enterprise's holistic design (gestalt, configuration, pattern) binding together its structure, strategy, environment and performance, showing its essential elements and the relationships among them, and mapping the interaction between the enterprise and is external environment, as both co-evolve over time" (Bozdogan 2007, p. 10). The EA is an abstraction of the "real-world" enterprise structured in such a way that makes both analysis and design possible by treating the enterprise as a hierarchical, nearly decomposable system. The EA has the potential to make the systematic design and improvement of complex enterprises more feasible by describing the enterprise from multiple perspectives and specifying the interactions of disparate aspects of the enterprise.

Although the original *raison d'être* for enterprise architecting was to integrate disparate information systems across the enterprise in order to provide value to the enterprise, its application has evolved from enterprise information system integration toward much broader uses. At its core, the purpose of enterprise architecting is (1) to understand the enterprise as an interacting complex system, (2) to communicate the structure and behavior to enterprise stakeholders, and (3) to serve as

---

* Many also hold that all enterprises can be said to have an enterprise architecture regardless of whether one was explicitly developed, as all enterprises can be described in terms consistent with an enterprise architecture.

a vehicle for changing the enterprise. It does this not so much by replicating the work done in other fields to understand key aspects of enterprises, but by harnessing them in a coordinated way: enterprise architecting "facilitate(s) the unification of methods of several disciplines used in the change process, such as methods of industrial engineering, management science, control engineering, communication and information technology, i.e., to allow their combined use, as opposed to segregated application" (Generalised Enterprise Reference Architecture and Methodology (GERAM) v. 1.6.3 1999). The documented EA can be used to foster a shared understanding of the enterprise's structure, function, and behaviors by enterprise stakeholders and can serve as the blueprint for changing the enterprise. When properly executed, enterprise architecting is a holistic integrated approach to the design of enterprises based on principles of near-decomposability and the facilitated application of many previously independent disciplines.

Despite its widespread application in industry and government, enterprise architecting has not emerged as a prominent field of academic study. The field has largely emerged from the practitioner literature rather than the academic literature, and as such, it has traditionally not had strong ties to rigorous theory despite its success and utility in practice. It does, however, offer one of the few practical approaches to integrating a vast array of knowledge about enterprises together with a holistic analytical approach that is able to address the fundamental complexity of enterprises. Some deficiencies in recent architecting practice that can be resolved using the ESE/A approach include the static nature of the descriptions and the difficulty in representing uncertainties in a dense set of data describing important features of the enterprise.

3.4.1 Enterprise Architecture Framework Views

Reference architectures provide a template for how enterprises can be decomposed in ways that make their complexity more manageable. The Generalized Enterprise Reference Architecture (GERA) establishes the use of *views* to accomplish the decomposition.* "Views contain a subset of facts present ... allowing the user to concentrate on relevant questions that the respective stakeholders may wish to consider. Different views may be made available highlighting certain aspects of the architecture and hiding all others." (ISO 15704 2000). A view in an enterprise framework is an abstraction that defines a perspective of an aspect of an enterprise, such as its organizational structure, its processes, or its information systems. The use of views in EA is analogous to the use of views in traditional building architecture: an architect will create a collection of drawings of a building from

---

* Although ISO 15704 specifies the use of the term "view," it is not universally used by all frameworks. Some frameworks use "viewpoints," "perspective," or "domain." Although each of these terms has its own particular definition, they are very similar concepts and date back to the Zachman Framework's use of domains and views.

multiple perspectives to provide stakeholders ranging from owners to contractors with a model of the building to be built. These perspectives may vary based on the location of an external observer (floor plan, street view, three-quarter view), or based on a functional set of perspective views (structural, electrical, plumbing, environmental system). Without any one of these views, not enough information is present to build the building. It takes the complete set. Unlike traditional architecting of buildings, however, enterprise architecting has no universally agreed-upon set of views that can completely specify an enterprise, or even how many views are required to describe it completely. An enterprise is considerably more complex than a building.

3.4.2 The Use of Views in Popular Enterprise Architecture Frameworks

There is no commonly agreed-upon number or set of views universally employed to describe EAs; EA frameworks currently in use have been developed based on best practice rather than a theoretical foundation. A representative set of views based on the literature is shown in Table 3.1.

Each framework employs a unique collection of views to describe the enterprise in a way that suits the objectives of the framework. Some frameworks will "matrix" views; for example, the Zachman Framework has six "viewpoints"[*] matrixed with six areas of focus to create a 36-cell framework of areas to be described, as shown in Figure 3.1. In another example of a matrixed architecture framework, both GERA and the computer integrated manufacturing open system architecture (CIMOSA) framework contain three dimensions of description: the views, the enterprise life cycle (identification/concept through decommission), and the level of specification (general, partial, and specific). The concept level of specification is common to all frameworks, but is specifically highlighted by CIMOSA. The generic specification is the structure given by the reference architecture; the partial specification is a reusable description that can still be adapted, and the specific is an architectural description specific to a particular enterprise.

There is no one "best" enterprise architecture framework that is best suited to all enterprises in all environments and for all applications. Each framework has its strengths and weaknesses, and is biased toward certain applications. As a result, it is not uncommon for enterprise architects to create "blended" frameworks that borrow best practices from a number of frameworks (Sessions 2007).

The Federal Enterprise Architecture (FEA) framework makes use of reference models (RM). Collectively, they are intended to provide universal definitions and constructs of the business performance, services, data, and technology elements of the enterprise and their alignment. This set of EA RMs serves as a foundation

---

[*] Enterprise architecture frameworks do not use a consistent terminology for the concepts of views.

Table 3.1 Views Employed by Popular Enterprise Architecture Frameworks

Enterprise Architecture Framework	Views Employed
Computer Integrated Manufacturing-Open System Architecture (CIMOSA)	Functional Information Resource Organization
Zachman Framework	Customer Owner Designer Builder Worker
Federal Enterprise Architecture (FEA)	Performance Reference Model Business Reference Model Data Reference Model Services Reference Model Technology Reference Model
The Open Group Architecture Framework (TOGAF)	Business Architecture Application Architecture Data Architecture Technical (Infrastructure) Architecture

to leverage existing processes, capabilities, components, and technologies as government organizations build target EAs. They are designed to facilitate cross-organizational analysis and the identification of duplicative investments, gaps, and opportunities for collaboration within and across federal agencies and other government organizations.

For a comparison of popular EA frameworks, including strengths, weaknesses, and suitability for different purposes, see Sessions (2007) and Schekkerman (2006).

3.4.3 Interactions Among Enterprise Architecture Views

Most EA frameworks currently in use either explicitly or implicitly assume that the EA can be described by considering the independent contributions of the set of views that it espouses. Zachman, when establishing the use of views in his framework, noted that "each of the different descriptions has been prepared for a different

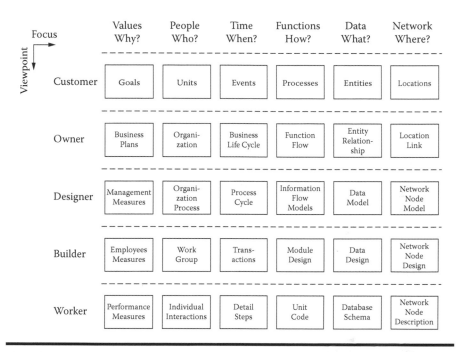

Focus Viewpoint	Values Why?	People Who?	Time When?	Functions How?	Data What?	Network Where?
Customer	Goals	Units	Events	Processes	Entities	Locations
Owner	Business Plans	Organization	Business Life Cycle	Function Flow	Entity Relationship	Location Link
Designer	Management Measures	Organization Process	Process Cycle	Information Flow Models	Data Model	Network Node Model
Builder	Employees Measures	Work Group	Transactions	Module Design	Data Design	Network Node Design
Worker	Performance Measures	Individual Interactions	Detail Steps	Unit Code	Database Schema	Network Node Description

Figure 3.1   A simplified Zachman Framework matrix. (Reprinted from http:// csrc.nist.gov/nissc/1996/papers/NISSC96/paper044/baltppr.pdf. As a work of the U.S. federal government, this image is in the public domain.)

reason, *each stands alone, and each is different from the others*" (emphasis added) (Zachman 1987, p. 282). Although Zachman notes that the descriptions are "inextricably related," he maintained that the documentation of each view must stand on its own.

There is a fundamental flaw with this approach to EA: the views of an EA are not independent. The views cannot be logically partitioned. In reality, each view is dependent on structures and behaviors in many other views, and overlap between views essentially always occurs. Behaviors and structures captured in any one view have cascading effects that cross the boundaries that exist between the definitions of views in any framework. For example, changes to an organizational structure do affect processes; changes to processes do affect knowledge requirements and information needs. Most users of architecture frameworks would agree that enterprises are complex, however, regarding adaptive systems with a high degree of interdependency between their components, too many then ignore many of the conceptual interdependencies that exist when creating or documenting an EA, despite the fact that the EA should be a tool to battle this very complexity.

As complex systems, enterprises are nearly decomposable, not fully decomposable. According to Simon's notion of near-decomposability, over the short term, the behavior of the subsystems will exhibit approximate independence. Over the long term, however, the behavior of any subsystem is dependent in an aggregate way on the behavior of the other components (Simon 1962).

The interactions between the major subsystems are significant drivers of enterprise behavior at these longer timescales and must be accounted for in any holistic analysis of system behavior. These views and the RMs provide structures and levels of integration that can be modeled more easily than traditional EA representations.

3.4.4 Application of Enterprise Architecture Frameworks in Systems Engineering

In contrast to the traditional architecture that creates blueprints for the construction of new buildings, enterprise architecting is most often used to support the efforts of enterprises already in operation that seek to change in some way. Enterprise architecting efforts begin with the "current state" EA: a blueprint of the enterprise in its current form. The current state architecture serves as the baseline for all enterprise change activities. In an ideal environment, an enterprise will always have its current state architecture updated and available.

In enterprise change efforts, some change has occurred that has made the current state architecture undesirable. Perhaps the environment has changed, rendering old processes inadequate; perhaps the enterprise's strategy has changed in order to become more competitive in the marketplace. When a change is desired from the current state architecture, the first step is to analyze the current state to identify where changes should be made to achieve the desired behavior for the transformed state. Currently, there are few rigorous processes for behavioral or performance analysis of EAs. In practice, many "analyses" consist of establishing carefully crafted narratives that link architecture to performance based on observation; the better analyses will involve an application of metrics to support the narrative. Value stream mapping (Rother and Shook 2003) and business process mapping (Sousa et al. 2002) are typical of the type of analysis conducted on EAs. Such activities can potentially be very useful and are accessible to many organizations but cannot provide a true analysis that links an enterprise's performance to its architecture or measure how a change to the architecture will affect enterprise performance. As complexity in the EA grows, the effectiveness of such static qualitative analysis techniques decreases.

Currently, the analysis of current state and future state EAs is a weakness of most enterprise architecting efforts. In some cases, analysis of the current state may be skipped entirely in favor of directly creating an alternative future state that addresses perceived problems. Such a weak treatment of analysis has the potential to completely undermine the value of enterprise architecting.

Despite the fact that EA frameworks represent an attempt to provide a systems thinking approach to describing architecture, if the architecture is not also analyzed using a systems thinking approach it may prove to be ineffective. As a complex system, changes made to one part of the architecture may have unintended consequences elsewhere in the system. A proper analysis of the EA will help us to understand where it can be decomposed and simplified and where it cannot. Without a rigorous approach that can analyze the EA and the behaviors it can potentially produce, the potential for unintended consequences is high and the utility of enterprise architecting as a management tool is almost negligible.

Some of the most promising approaches for analysis of EAs employ simulation techniques that model the behavior of the enterprise over time using the information captured by the EA. The vast majority of attempts to simulate EAs have focused on simulating information systems, as EA has disproportionately been focused on information system efforts and it is more straightforward to model the logical nature of information systems when compared to other aspects of the EA, such as organizational behavior or knowledge.

3.4.5 Integration with Enterprise Systems Engineering and Simulation Models

EA frameworks, even with the inclusion of the perspectives and guidance of organizational theory, do not provide a complete toolset for architecting the enterprise. EA frameworks have the potential to provide a series of lenses, theories, and tools for reducing the apparent complexity of enterprises. A good framework decomposes the enterprise into manageable views that can be interpreted and understood both independently and in relation to other views. It does not, however, necessarily help managers understand how the EA will behave in different environments with different inputs, or if the architecture will enable or preclude an expected level of performance.

An EA created using a framework is a static description of the essential components of the enterprise and their interconnections. By itself, this static description does not provide enough information to analyze and understand the behaviors that a given EA is capable of producing. Enterprise managers need the analytical capabilities presented here—enterprise systems engineering including optimization and simulation modeling, and coupled with the framework for architecting—to help analyze, understand, and manage transformation activities. As noted above, analytical capabilities are essential to the development of a true process for enterprise architecting (Levitt 2004). Without such capabilities, there can be little assurance that any changes to the architecture will have their desired effect or that any understanding of enterprise behavior is correct.

Later chapters address these analytic tools by exploring the literature to identify appropriate simulation methodologies that can be brought to bear as illustrated in Section II. Using the decomposition of enterprise frameworks and organizational science as guides, the contribution of different simulation perspectives is explored, and a hybrid approach to simulation modeling is developed that seeks to combine these perspectives in a way that is able to harness the strengths of these simulation methodologies.

3.5 Conclusion

There is a strong relationship between enterprise architecting as has been practiced in the private sector and government, and the dynamic methods employed in enterprise systems engineering. This chapter outlined the formulation and experience with these methods and their integration in ESE/A with an emphasis on the dynamic behaviors of the enterprise and its transformation to adapt to the evolving external environment. This foundational ESE/A process advances enterprise architecting from a static and deterministic regime to a dynamic and probabilistic regime.

The case studies in Section II demonstrate a mature state of development in the dynamic modeling of enterprise performance in operations, supporting information flows for planning and operational management, and the costs and benefits of transforming systems and technologies. Because of the critical importance of aligning organizational change with the transformation of the enterprise, attention has been given in this chapter to the state of development of organizational science and practical approaches to advancing that state.

The foundation established here for the ESE/A process is expanded upon in the following chapters of Section I that review specific models and methods to describe enterprise dynamics (Chapter 4) and that present a management approach that incorporates ESE/A supported by enterprise dynamics methods to guide the transformation process. Emphasis is placed on governance in Chapter 5 to assure performance of all required roles and responsibilities in a coordinated way.

References

Anderson, P. et al. 1999. Introduction to the special issue: Applications of complexity theory to organizational science. *Organization Science,* 10(3): 233–236.

ANSI/IEEE 1471:2000. IEEE Recommended Practice for Architectural Description of Software-Intensive Systems.

Bernus, P. 2003. Enterprise models for enterprise architecture and ISO9000:2000. *Annual Reviews in Control,* 27: 211–220.

Bozdogan, K. 2007. Enterprise architecture modeling, design and transformation: Defining the missing links, *Lean Aerospace Initiative,* Cambridge, MA: MIT, April 2007.

Carley, K.M. 1995. Computational and mathematical organization theory: Perspective and directions. *Computational and Mathematical Organization Theory,* 1(1): 39–56.

Cyert, R. and March, J. 1963. *Behavioral Theory of the Firm.* Oxford: Blackwell.

Ethiraj, S.K. and Levinthal, D. 2004. Modularity and innovation in complex systems. *Management Science,* 50(2): 159–173.

Fowler, A. 2003. Systems modelling, simulation, and the dynamics of strategy. *Journal of Business Research,* 56(2):135–144.

Galbraith, J.R. 1973. *Designing Complex Organizations.* Reading, MA: Addison-Wesley.

GERAM Generalized Enterprise-Reference Architecture and Methodologies Appendix B ISO 15704 standard.

Glazner, C.G. 2009. Understanding enterprise behavior using hybrid simulation of enterprise architecture. Dissertation. Cambridge, MA: MIT.

ISO 15704 2000. Industrial automation systems—Requirements for enterprise-reference architectures and methodologies.

Levitt, R. 2004. Computation modeling of organizations comes of age. *Computational & Mathematical Organization Theory,* 10: 127–145.

March J. and Simon H.A., 1958. *Organizations.* New York: Wiley.

Milgrom, P. and Roberts, J. 1995. Complementarities and fit strategy, structure, and organizational change in manufacturing. *Journal of Accounting and Economics* 19: 179–208.

Mingers, J. and Gill. A. 1997. *Multimethodology: Towards a Theory and Practice of Combining Management Science Methodologies.* Chichester, UK: John Wiley & Sons.

Orton J. and Weick, K. 1990. Loosely coupled systems: A reconceptualization. *Academy of Management Review, 1990,* 15(2): 203–223.

Rivkin, J.W. and Sigglekow, N. 2003. Balancing search and stability: Interdependencies among elements of organizational design. *Management Science* 49(3): 290–311.

Rother, M. and Shook, J. 2003. *Learning to See: Value Stream Mapping to Create Value and Eliminate Muda,* Cambridge, MA: Lean Enterprise Institute.

Schekkerman, J. 2006. *How to Survive in the Jungle of Enterprise Architecture Frameworks.* Victoria, BC: Trafford.

Scott, W. and Davis, G. 2006. *Organizations and Organizing: Rational, Natural, and Open Systems Perspectives.* Upper Saddle River, NJ: Prentice Hall.

Sessions, R. (2007). Building distributed applications—A comparison of the top four enterprise-architecture methodologies. Microsoft Developer Network. http://msdn.microsoft.com/en-us/library/bb466232.aspx (Accessed April 17, 2012).

Simon, H.A. 1962. The architecture of complexity. *Proceedings of the American Philosophical Society,* 106(6, (Dec. 12): 467–482.

Sousa, G.W.L. et al. 2002. Applying an enterprise engineering approach to engineering work: A focus on business process modeling. *Engineering Management Journal,* 14(3): 15–24.

Swinth, R.L. 1974. *Organizational Systems for Management: Designing, Planning, and Implementation.* Columbus OH: Grid Inc.

Tatum, C.B. 1983. Decision-making in structuring construction project organizations. In *Technical Reports of the Department of Civil Engineering.* Vol. 279, Stanford, CA: Stanford University.

Thompson, J. 1967 *Organizations in Action: Social Science Bases of Administrative Theory.* New York: McGraw-Hill.

von Bertalanffy, L. 1956. An essay on the relativity of categories. *Philosophy of Science,* 22(4): 243–263. Chicago: University of Chicago Press.

Wiener, N. 1956. *I Am a Mathematician: The Later Life of a Prodigy.* Garden City, NY: Doubleday.

Zachman, J.A. 1987. A framework for information systems architecture. *IBM Systems Journal* 26(3): 276–292.

Chapter 4

Enterprise Dynamics Methods and Models

Kenneth C. Hoffman, William J. Bunting,
Christopher G. Glazner, and Leonard A. Wojcik

Contents

4.1 Introduction

Multiple dimensions of complex large-scale enterprises—including number of stakeholders, emergent properties, changing environments and landscapes, and uncertain futures—pose enormous technical challenges requiring the contributions of multiple disciplines, theories, and methods. These multiple dimensions are categorized here using the policies, organizations, economics, and technology (POET) concept covering policy, organization, economics, and technology. Describing and analyzing the dynamic interactions across these POET dimensions in the operations and transformation of complex enterprises are the principle objectives of enterprise dynamics.

This chapter summarizes a diverse set of models and methods that can be drawn upon to address specific dynamic aspects of the enterprise as functional source modules. Several have been applied in the enterprise systems engineering and architecting (ESE/A) case studies presented in Section II of this sourcebook. They represent differing perspectives across disciplines and are drawn from an even larger population of methods that are used for similar analytical purposes. The dynamics of a multiagency mission or undertaking (a multiagency enterprise) can best be understood as a complex system-of-systems (SoS) amenable to modeling and simulation.

Because most models and, indeed, disciplines describe partial, albeit important, aspects of an enterprise, there is a need to integrate or more strongly unify them to provide the more comprehensive perspective required to plan and manage enterprisewide transformation. The analytic concept of a unifying framework that integrates features of the diverse and specialized source modules (models) into a comprehensive state-space representation of the dynamic enterprise is formulated in this chapter; its feasibility is demonstrated in a Section II case study, Chapter 8, to analyze the effectiveness of the systems engineering (SE) process on a major transformation program.

The following interrelated enterprise dynamics topics are described further in major sections below:

- Enterprise landscape representations that capture the many state-space dimensions and the complex characteristics and interacting elements of the enterprise derived from and related to the enterprise architecture
- Models and methods providing functional source modules for enterprise dynamics drawn from multiple disciplines to address specific aspects of the state-space and dynamic behaviors across the landscape
- Methods to deal with uncertainties in enterprise transformation using Bayesian and line of sight concepts
- Policy analysis and technology assessments for diverse and rugged enterprise landscapes using a simplified illustrative example
- Unifying approach to enterprise dynamics that enables comprehensive dynamic analyses within the ESE/A process by integrating a diverse set of models for a more comprehensive state-space description of the enterprise and its transformation path

Enterprise planning and analysis in the public and private sectors focus on operational performance of the mission or business objectives, organizational effectiveness, acquisition of technology for new capabilities and a strategic edge, and the management and decision-making regimes leading to successful implementations. No single modeling method can capture these complexities and range of applications; hence again the emphasis here is on unifying approaches. The term "unifying framework" is chosen to emphasize the ongoing process of interpreting, assimilating, and, where appropriate, integrating the contributions of diverse disciplines and models to inform acquisition and operations managers rather than the unrealistic end state of a "unified theory" grand or otherwise.

State-space descriptors in the unifying framework include operational parameters such as mission performance levels, resource requirements, organizational structures, information exchanges, personnel skill levels and turnover and training levels, and environmental parameters describing the market or mission space and adversarial or competitive forces. The progress and outcomes of acquisition programs to develop improved capabilities are represented and involve the level of technical and process change, pressures to change, and costs.

The unifying framework may be formulated as a nonlinear, or piecewise linear, control theory representation of state-space elements that can be controlled and influenced to varying degrees, or that are beyond the control of the enterprise but require adaptive sense-and-response mechanisms, domains that are applicable to the enterprise at any scale. Specific modeling methodologies that can be used to describe selected dynamic elements of the enterprise state-space include systems dynamics, Fourier analysis, and agent-based and highly optimized tolerance methods.

4.2 Enterprise Landscapes, Perspectives, and Elements

Descriptive enterprise dynamics builds on the concept of descriptive geometries and involves the mapping of the highly multidimensional enterprise state-space into the decision space for management of system and technology acquisition, policy formulation, and business operations.

The objective of research on descriptive enterprise dynamics is to develop and apply dynamic theory to acquiring and implementing technology, systems, and services for new operating capabilities in very complex organizational and operational environments. Understanding the characteristics of the enterprise landscape and support for specific management decisions will increase the opportunities for success in this failure-prone endeavor.

4.2.1 Enterprise Landscapes

The concept of landscapes has been applied to organization theory and enterprise fitness (Levinthal 2001) to deal with operations performance and management regimes (Snowden and Stanbridge 2005) that deal with people, organizations, and policies. It has also been applied to various scientific domains, including biological evolution (Crutchfield 2003) and string theory in physics at a cosmological scale (Bousso and Polchinski 2004). Enterprise landscapes are used here as a basic structure for descriptive enterprise dynamics (Hoffman et al. 2007).

The enterprise fitness landscape in Figure 4.1 displays the characteristics of an enterprise in terms of its near- and long-term responses to a changing external environment. It offers a high-level aggregate view of the state-space of the enterprise. Architecting the enterprise structure is an effective method of planning and managing change in all quadrants, but is applied with a different perspective depending on the specific challenges.

The enterprise management landscape is shown in Figure 4.2. It provides a complementary perspective using a two-dimensional matrix. One dimension represents repetitive business processes (smooth landscape) versus ad hoc reactive/proactive activities (rugged landscape). The other dimension represents the decision domains of rules-based versus reactive situation dependency.

Specific modeling approaches and state-space descriptors are highly dependent on the position of the enterprise on this landscape. This sourcebook emphasizes the upper-right social complexity domain involving ad hoc operations (rugged landscape) and situationally dependent decisions, although other regimes of the matrix are addressed as well by tailoring and perhaps simplifying the methods. Much of

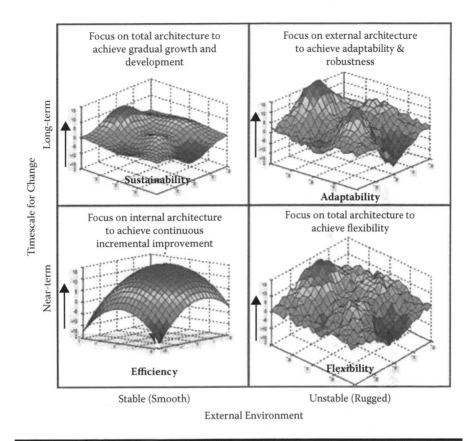

Figure 4.1 The enterprise fitness landscape. (Reprinted from K. Bozdogan, C. Glazner, K. Hoffman, and J. Sussman, Lean Aerospace Initiative Annual Conference, Boston, MA: MIT, 2008.)

the modeling and simulation literature deals with the lower-left quadrant, process engineering of highly repetitive and rule-based enterprise operations, and is not readily adaptable to the social complexity domain characterized by emergent behavior and complex adaptive systems theory.

In addition to position on the management landscape, analytical methods are also dependent on the operational, scalar, and temporal perspectives of interest in specific applications. Descriptions of these perspectives follow. The operational perspective provides a template for the core approach; the hierarchy and importance of the other perspectives is problem dependent.

4.2.2 Core Operational Perspective

This work uses the operational domain as the core perspective for enterprise dynamics as it is this perspective that drives major acquisitions to improve operational

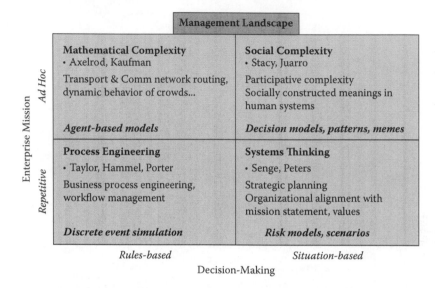

Figure 4.2   The enterprise management landscape. (Adapted from D. Snowden and P. Stanbridge, *E:CO* 6(1–2): 140–148, 2005.)

capabilities. The operational perspective represents the enterprise in terms of controlled, influenced, and uncontrolled (C-I-U) spaces and applies to enterprises of any scale at any stage in the temporal life cycle. This operational perspective (DeRosa 2005) is illustrated in Figure 4.3.

Sense-and-respond decision-making processes are mapped into these domains and are key factors in the dynamic behavior of enterprises. The content of the C-I-U domains, uncertainties, and the sense-and-respond mechanisms including their nature and timing, vary across the domains as a function of the "smoothness or ruggedness" of the management landscape for specific enterprises.

4.2.3 Scalar Perspectives

Scalar perspectives of the enterprise are defined in terms of both organizational and enterprise scale:

- *Organizational Scale.* An organization can be an individual, a unit, a bureau, an agency, or be multiagency. This chapter focuses on multiagency organizations performing complex missions where the greater understanding of dynamic interactions is most urgent.
- *Enterprise Scale.* Enterprise perspectives can be taken at multiple scales. The complex, large-scale, integrated, open-systems (CLIOS) method discussed in a later chapter provides a unifying approach to multiscale analysis applied to transportation planning across social, economic, technical, and policy dimensions.

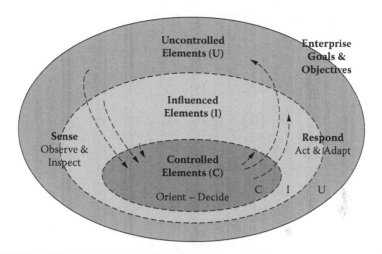

Figure 4.3 Operational perspective defining the enterprise in terms of controlled, influenced, and uncontrolled domains: the basis of the core C-I-U enterprise model. (Adapted from J.K. DeRosa, Presentation to the Faculty of the College of Engineering and Mathematical Sciences at the University of Vermont, Burlington. December 2005.)

In this sourcebook, we address enterprise dynamics on the following enterprise scales:

■ *Acquisition Enterprise* focuses on the stakeholders involved in major technology acquisition programs, for example, a global information grid, or nationwide health information network. Real options concepts (Dahlgren 2007) address improvements to acquisition strategy that can add long-term flexibility and adaptability to the enterprise. Most transformations involve modernized information systems and software, a high-risk area. The ESE/A process incorporates software and information systems in the context of enterprise operations to ensure that requirements are related to real performance enhancements. The details of software management are beyond the scope of this sourcebook and managers should draw upon the extensive literature and experience on this topic.

■ *Operational Enterprise* focuses on the agencies and agents (public and private), managers, and staff engaged in multiagency mission or business operations, for example, health services, security operations, and tax processing.

■ *Economic Sector* focuses on the activities performed in a large sector of the U.S. economy, for example, federal health programs or air traffic management.

Each of the above enterprise scale categories involves individuals, units, and larger organizations with multiple stakeholders internal and external to the enterprise.

4.2.4 Temporal Perspective

The acquisition and transition to operations life cycle provides the temporal perspective. This life-cycle perspective poses major management and decision-making challenges in an enterprise that is acquiring new operational capabilities: the focus of the ESE/A process.

Specific steps in the temporal (life-cycle) perspective include:

- Research and development
- Technological systems
- Organization and social systems
- Policies
- Planning and acquisition strategies
- Strategic planning
- Requirements and capability planning
- Contractual acquisition, systems/services
- Transition to operations
- Business process change
- Organizational change
- Training
- Implementation and deployment strategy
- Operations
- Performance measurement
- Ongoing system or service enhancement

The failure rate of major acquisitions is unacceptably high—over 50% for information technology (IT) projects (Flint 2005)—and motivates enterprise dynamics toward a better understanding of causal factors in those failures. The sense-and-respond elements of the enterprise, as represented in the operational perspective, are critical to the success of major acquisitions.

4.2.5 Elements of Enterprise Transformation Dynamics and Complexity—The ESE Profiler

The dynamic processes that affect the acquisition and implementation of new capabilities in complex enterprises include:

Performance Dynamics in acquisition programs and in ad hoc proactive and reactive missions and business initiatives

Organizational and Behavioral Dynamics involving collaboration across agencies and among stakeholders to perform complex acquisitions and missions

Information Dynamics to collect and exchange information that supports sense-and-respond decision processes during acquisition and the transition to operations

Technology Dynamics governing the maturity, rate, and degree of difficulty in employing advanced technologies for improved enterprise performance

The content and importance of these dynamic elements vary depending on the enterprise landscape and perspective of interest. There is extensive literature in each of these areas of enterprise dynamics. The complexity of enterprise transformation programs varies considerably based on the size and scope of the enterprise, desired capabilities, and strategies. These complexity metrics may be mapped using the Enterprise Systems Engineering Profiler developed at MITRE (Stevens 2010): "The Systems Engineering Profiler is intended as a first step toward the development of a self-assessment tool that can help the systems engineer understand the nature and context of the system of interest. It is also intended as the basis of a situational model that would help systems engineers select and adapt the processes, tools, and techniques most applicable to the particular system problem and context."

As a self-assessment tool, the Profiler can help the systems engineer understand the nature and context of the "system" of interest and the context in which it will be developed and will operate. In effect, the manager or engineer can use this framework to map the system or megasystem of interest, creating a spider chart or polar diagram of the system's context.

As a situational model, the Profiler can help the system engineer select and adapt the best processes, tools, and techniques on the basis of the system's nature and context. Underlying the very notion of a situational model is the premise that different processes, tools, and techniques apply in different situations. The challenge is to understand the situation sufficiently well to select the most appropriate tools and adapt them as the situation warrants. The Profiler is organized into four quadrants and three rings. The quadrants describe different dimensions of the broader context in which the system or megasystem will be developed, will operate and evolve. The three concentric rings reflect increasing levels of complexity and uncertainty.

Figure 4.4 uses the Profiler to portray complexity in terms of the four major quadrants (Strategic, System, Stakeholder, and Implementation contexts). Each of the four quadrants is, in turn, further decomposed into two related dimensions resulting in eight dimensions. Each dimension ranges from relatively simple patterns in the inner ring, to highly complex characteristics in the outer ring. By applying the Profiler early in the planning cycle, all stakeholders can identify their perspectives on the scope and complexity of a transformation plan and expose inherent risks. This step supports a rigorous comparison of the risks of transformation against the expected benefits.

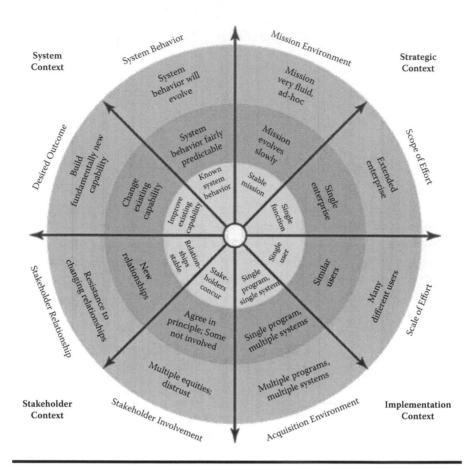

Figure 4.4   Enterprise Systems Engineering Profiler.

4.3 Representative Methods for Enterprise Dynamics

A diverse set of methods is applied in the source modules within the unifying framework. *Source modules* are defined here as models based on specific methodologies that are applied to selected elements of enterprise dynamics. Although the methodologies are often related to a specific discipline (e.g., engineering, economics, biology, policy science), they generally take a multidisciplinary approach when applied to complex systems. A variety of such source modules is being studied to address specific acquisition challenges from the appropriate perspective and at the required level of fidelity. They address various stages in the temporal life cycle at differing scales, or level of fidelity.

In the following sections, information flows from the source modules are described through the unifying state-space descriptors that may be derived, to the concept of the C-I-U enterprise model. The models selected for inclusion in this initial version of the unifying framework and their roles are described below.

The methods deal with uncertainties in both an implicit and explicit fashion. The Bayesian method is particularly powerful in this respect, and the illustrative unifying framework can deal explicitly with the effect of uncertainties on acquisition strategy and enterprise applications.

4.3.1 Systems Dynamics Risk Model

Systems dynamics (SD) (Love 2006) had its origins in control theory and can represent a rich set of influences on the enterprise with feedback relationships. An SD model was applied in combination with a materials flow model to capture dynamic influences within nuclear waste management. This SD model was then applied to evaluate alternative strategies for the processing of nuclear waste from defense programs and is described in a case study, Chapter 10. An SD risk model has also been developed to project the evolution of a transformation or acquisition program based on requirements stability, technology maturity, and program complexity. This model captures feedback and feedforward effects that influence progress and cost.

Specific questions that may be addressed with the models include:

- What are the requirements of transformation or acquisition to support enterprise operations objectives?
- What are the influences of new technologies and the changing external environment on requirements, transformation strategies, and capabilities?
- What level of performance can be achieved given the plan and resources guided by critical decisions?

4.3.2 Agent-Based Acquisition Stakeholder Model

An agent-based model (Mathieu et al. 2006) can identify roles and actions of stakeholders (agents) and groups (more than one individual) in a major transformation program. In Chapter 9, agent-based modeling has been applied in a hybrid formulation to the simulation of an operations center in a case study. In transformation programs, the stakeholders and groups interact and make decisions during the various life-cycle phases (e.g., milestone reviews, technology development, etc.), and are subject to different environmental pressures (e.g., budget cuts, oversight, and public perception). In its current version, the agent-based model is linked to the SD acquisition process model: decisions are informed by running the agent-based model and delays are determined by running the SD model.

Specific questions that may be addressed with the agent-based model include:

- What are the influences of stakeholder interactions and decisions on the acquisition process?
- What are the influences of the changing policy environment on stakeholder interaction and decisions?

4.3.3 Highly Optimized Tolerance Model

Highly optimized tolerance (HOT) (Wojcik 2004) is a framework for understanding certain aspects of complexity in designed, engineered, and managed systems. Carlson and Doyle (1999) originally created the HOT concept and applied it to forest fire management, Internet traffic, and natural ecologies. For engineered systems, they showed how power laws relating event size to probability emerge from minimization of expected cost in the face of design trade-offs and uncertainty. Since then, the HOT methodology has been applied to such systems as the electric power grid and Internet architecture. HOT typically is identified with power laws. The HOT model is applied in the case study in Chapter 8 to model the systems engineering process applied to a specific enterprise transformation program.

Specific questions that may be addressed with the HOT model include:

- How is the pace of a transformation program influenced by cost and management pressures?
- What is the expected transformation progress to be achieved given the schedule of planning, budgeting, and decision cycles?
- What are the influences of alternative engineering processes and acquisition strategies?

4.3.4 Technoeconomic Model

This modeling approach describes an operational enterprise environment in sectors of the economy influenced by new technologies and public preferences. It links systems engineering process models of technologies and systems providing major services in sectors of the economy represented in an interindustry macroeconomic model. The approach has been applied to technoeconomic analysis in the energy sector (Jorgenson 1998) and is being adapted to analyze the potential of technologies to improve health services and their impact on the 16% of the gross domestic product (GDP) that is devoted to the healthcare sector of the U.S. economy.

The technoeconomic modeling approach is applied to resource and energy planning in the Chapter 12 case study and to the analysis of future trends in healthcare in the case study in Chapter 13.

Specific questions that may be addressed with the linked systems engineering and socioeconomic model include:

- How can technology be introduced most effectively into business processes and services?
- What are the impacts of major technological and policy changes on the growth of the enterprise in its economic sector?
- What are the multilevel impacts on cost-of-service and employment in the economic sectors involved?

4.3.5 Cycles and Lags Acquisition Field Effects Model

The cycles and lag model (Rahimzadegan, West, and Hoffman 2006) applies Fourier series to project progress in an acquisition program as influenced by the periodic planning, budgeting, and review cycles dictated by acquisition governance polices as well as common programmatic issues that arise periodically. These "field effects" are dictated by governance policies and must be overcome in all programs. Separate models may be incorporated that describe the influence of organizational factors and program complexities that influence progress.

Specific questions that may be addressed with the cycles and lags model include:

- What is the expected acquisition progress to be achieved given the annual planning and budgeting cycles dictated by governance policy, the normal personnel change cycles, and the typical lag between occurrence of a problem and its resolution
- What are the influences of alternative acquisition strategies, for example, prime or large-scale integrator, managed services, and venture model?
- What are the influences of organizational features and program complexity?

4.3.6 Homeodynamic Risk–Opportunity Model

The homeodynamics model uses fluidics and biological analogues to simulate resistance to change in enterprise and response mechanisms. The method derives from biological systems (Yates 2002) and may be applied to enterprise operations where an external parameter λ applies to the dynamic influence of the external environment on the enterprise (analogous to a shear "viscosity"). An internal parameter η represents the responsiveness of the enterprise to the boundary effect (analogous to a bulk "viscosity"). White (2006) describes an application of the concept to a dynamic enterprise affected by risk while responding to opportunities.

Specific questions that may be addressed with the homeodynamics model include:

- What is the expected acquisition progress to be achieved given the rate of internal enterprise responses to changes in the external environment?
- What are the natural risk–opportunity trade-off dynamics that arise in programs?

4.3.7 Line of Sight Evidential Reasoning Analysis

In order to examine performance dynamics of complex enterprises with consideration of the uncertainties inherent in enterprise transformation, it is necessary to apply reasoning methods along with simulation and modeling methods. Enterprise transformation is a complex undertaking involving organizational change and possibly new technology utilization. Chief executive officers (CEOs) require reasoned alignments of the enterprise's technology investments and performance goals, and need to understand how the technology investments support the attainment of the transformation. These reasoned alignments require making complex arguments that integrate information about strategic planning, enterprise modeling, enterprise dynamics, and SE.

Line of sight evidential reasoning analysis (LSERA) (Bunting 2012) is a method for reasoning about the likelihood of attaining specified desired performance dynamics resulting from emerging dynamics and uncertainties within an enterprise or its environment. A basic principle is that mission performance occurs from individuals understanding the dynamic interactions of people, process, and technologies. Individuals reason on dynamic interactions, producing insights on likely performance attainment given the interactions. Individuals increase their respective understanding and their subsequent decisions and actions are changed because of their improved collective understanding.

The method uses evidential reasoning theory (Schum 1994) supported by multi-entity Bayesian network (MEBN) theory (Laskey 2008) and represents changes in dynamics by adding "sensed" information and actions (interventions) as nodes to the MEBN network. The method supports reasoning around the enterprise line of sight (ELoS). A technology initiative's ELoS is an inferential alignment of a technology initiative to business processes, business services, and strategic outcomes, supporting a principled assertion of attainment of the strategic outcome. The ELoS supports managers understanding, at all levels within the enterprise, the attainment of desired behaviors that in combination contribute to organizational performance dynamics and mission performance attainment.

LSERA supports reasoning from evidence to strategic outcome attainment assertions. Specific questions that the LSERA analysis may address include:

- If the enterprise invests in a given organizational and process change or technology initiative, will the enterprise attain its desired business results?
- How do uncertainties and perceptions bear on behavioral and decision processes?
- What is the change in the current likelihood of desired results attainment based on new emergent information?
- If management takes this action (intervention) to change the dynamics of the organization, will it increase or decrease the likelihood of desired results attainment?

LSERA consists of two main models: the argument model that structures the reasoning and evidence into a chain of if–then statements and the inference model that creates a Bayesian network probability model indicating the likelihood of a behavior attainment. Using these two models allows managers to reason about and gain insight from complex people, process, and technology interactions where enterprise analysts cannot easily construct mathematical algorithms. LSERA supports managers reasoning asynchronously within their particular areas and the results being analytically joined in a mathematically principled manner. Chapter 7 further explains these models along with an example.

4.3.8 Bayesian Networks

Uncertainties regarding changes to the enterprise and its operational environment and transformation may be dealt with implicitly or explicitly in planning and management. In particular, given the uncertainty within the enterprise landscape, managers and analysts often use probabilistic methods to reason given the results of other analytical techniques. Formal methods for dealing with uncertainties that can inform or validate experience are needed and Bayesian networks provide a powerful approach. Estimates of uncertainty are often drawn from the experience of managers, analysts, and other experts. Bayesian networks and methods address the need to deal explicitly with such uncertainties in enterprise transformation.

Bayesian networks represent the probabilistic relations among variables that collectively form a joint probability distribution. The variable will have states that are mutually exclusive (must be in only one state) and collectively exhaustive (there are no other states), for example, {true or false}, {male, female}. There are relations among the variables with direction of influence. Arcs depict the directed influences among the variables where the path (sequence of arcs) from a variable does not cycle back to the variable. Probabilities represent the influence strength of a variable's state on another variable's states. The existence or absence of arcs among the variables represents variable dependence/independence. An independence condition for a Bayesian network is that every variable is independent of its nondescendants given its parents.

Modeling enterprise dynamics can also have significant degrees of uncertainty. Bayesian network theory can be used for managing and understanding uncertainties within the enterprise dynamics and encoding the knowledge structure of the enterprise dynamics for inference and examination.

4.4 Coupling Enterprise Dynamics Models with Enterprise Architecting—ESE/A Process

The enterprise dynamics models described above provide an enterprise systems engineering (ESE) capability for transformation planning and management. This ESE capability resolves a major deficiency in the formerly static approach to enterprise architecting as was established in the foundational Chapter 3. The resulting method, ESE/A, provides a dynamic process to guide managers through the complexities of enterprise transformation. This method is sufficiently mature for current transformation applications as outlined in this section. In view of the rapidly escalating challenges and the research activity in this field, continued evolution of the ESE/A process is expected. A concept for this future of ESE/A is offered in Section 4.5.

4.4.1 Dynamic Enterprise Architecting Using Simulation Models

One promising approach to better understanding and analyzing enterprise dynamics is to simulate the behavior of an enterprise as guided by enterprise architecting. This approach holds that the behavior of enterprises can potentially be understood through simulation. The inherent complexity of enterprises, however, makes practical simulation a challenge. To address this complexity, enterprise architecture (EA) frameworks are used to decompose the enterprise into coherent views which can then be simulated. By integrating simulations of multiple EA views, a deeper understanding of dynamics driven by highly interdependent views and contexts can be built.

The subject of coupling simulation methods with EA (or architecting as the active form) was presented in Chapter 3. An application case study is presented in Chapter 6 with emphasis on its role in management decisions regarding the investment of discretionary funds. The architecting capability is enhanced significantly using simulation methods that capture the dynamic features of the POET elements of the enterprise.

4.4.1.1 Principles of Hybrid Simulation Models of the Enterprise Architecture

Simulation models have been used for a wide range of applications, including forecasting, education, and theory building, and as decision aids. Each application requires a different approach to model building and use as each application has different objectives and goals. A simulation model intended to predict the price of a traded commodity, for example, requires a different approach for creation and use than a simulation model intended to understand the spread of a contagious disease or one designed to teach people how to use a system. The guidelines for creating a simulation model change depending on whether the model is intended to describe, predict, optimize, build theory, teach, communicate, or uncover hidden mechanisms. The goals of simulation models espoused here are focused on understanding and communicating the effects of the enterprise's architecture on the behavior of the enterprise with enterprise leaders as the key model users. These simulation models must yield insight into complexity, while communicating and educating. They do not need to be predictive but they must be able to identify possible enterprise behavioral and performance outcomes in response to discrete strategic management decisions given its defined architecture. These simulation models must work well with existing abstractions and integrate disparate perspectives into a single hybrid simulation model. To meet these goals, a set of guiding principles has been developed to help guide the modeling process. There are four key principles that can be identified to aid the modeler in creating hybrid simulation models of an EA:

1. Models should be created for insight (e.g., through what–if analysis by defining possible future outcomes related to decisions) not for generating point predictions.
2. Models must capture the essential elements of the enterprise architecture and the architecture should contain source data for analytics.
3. Hybrid models should preferably be focused at the strategic level versus the tactical level to address enterprise-level strategic decisions and their possible consequences.
4. Hybrid models must explicitly capture interactions across the enterprise's architecture comprising multiple views or domains as required by the strategic decision question or issue being posed.

The next four sections develop these principles further for application to the creation of hybrid simulation models of EA.

4.4.1.2 Modeling for Insight, Not Prediction

EA simulation modeling is not a predictive approach to simulation modeling that is intended to improve operational efficiency; instead, this approach seeks to yield insight into the behavior of a complex enterprise arising from its architecture and the key interactions and decision points. This insight helps to accelerate the learning curve for enterprise managers seeking to shape and guide their enterprise from a system-level perspective. This approach allows the modeler and architect to capture key attributes of the enterprise from multiple perspectives (e.g., strategy, process, organization, products, etc.) and to examine how the interactions between these perspectives drive the high-level behaviors of the enterprise. Such an approach takes a strategic view of the enterprise to guide enterprise architects and managers in understanding how the system, as a whole, delivers value to its stakeholders. This use of simulation modeling as a strategic decision support tool is similar to that of researchers in the fields of systems thinking and organizational learning. Here, models are used along with architectures to gain insight into the system by:

1. Creating a shared frame of reference among managers and architects for understanding nonlinear enterprise dynamics
2. Understanding the relationship and effects between both hard (quantitative) and soft (qualitative) system variables
3. Testing hypotheses and performing scenario analyses
4. Discovering new architectures or ways to manage the current architecture as a result of analysis of the simulation

The last three points correspond closely to the three stages of learning espoused by researchers in the field of organizational learning (Argyris and Schön 1978; Senge 1990; Sterman 2000; Fowler 2003).

4.4.1.3 Strategic Level Modeling of the Enterprise Architecture

A key tenet of the proposed approach to hybrid modeling of EA is the focus on enterprise dynamics at a strategic level as opposed to tactical or operational levels. This has profound implications for the modeling approach: instead of a detailed model with high precision that aims for predictive capability in a single context, this approach aims to deliver a model that, although lacking high precision, delivers insight about the system in a far broader context in the face of uncertainty. Such an approach to modeling addresses fundamentally different questions from those addressed by tactical models. A tactical model might answer the question of what the parameters are that will allow achievement of optimum output, however, a strategic model will answer the question of what the design characteristics are that provide for good performance in the anticipated environment.

Although organizational design theorists, systems thinkers, and enterprise architects (Galbraith 1973; Sterman 2000; Ross et al. 2006) have embraced a strategic focus in the study of enterprises, this has not been reflected in the work of enterprise modelers. Most modelers tend to focus models either too narrowly (such as to support a single view in an EA), too tactically (as in the study of supply chain efficiency), or both (Kalpic and Bernus 2002; Epstein 2003; Schieritz and Größler 2003). Researchers in the field of computational organizational design have created models with a strategic focus (Rivkin and Sigglekow 2003; Sigglekow and Levinthal 2003; Ethiraj and Levinthal 2004). However, these models tend to be theory generative and too abstract to provide insight that most enterprise managers can understand and trust. There is a tremendous need for a strategy-focused modeling capability that will not only support the theories of organizational design theorists and systems theorists, but will also support the needs of enterprise managers as they seek to guide the development and direction of their enterprise.

4.4.1.4 Focus on Dynamics Resulting from Interactions Across the Architecture

The proposed approach to simulation modeling of EA seeks to capture the complex dynamics of the enterprise by explicitly capturing the interactions across the contextual boundaries of the EA's views in accordance with the concept of near-decomposability of complex systems. These interactions provide pathways for feedback in the system, both within and across the views of the EA. Beginning with a set of initial conditions, such a simulation model will play out the response of the EA over time to an external environment.

4.4.1.5 Process for Developing Hybrid Models of the Enterprise

Simulation models are used as part of an iterative process to help understand problems with the current architecture and provide input for creating new to-be architectures. This process helps ensure that the simulation model is properly bounded and scoped, and that a framing question is well posed, the model is created in a logical progression, and the model can be useful in answering the question it was intended to answer. Modeling will always be a creative endeavor, however, it can benefit from a structured process. Without a well-developed process to guide the model creation effort, the model can quickly become unmanageable or unsuitable to its purpose as modelers lose sight of model scoping, purpose, or structure. A process that is standardized helps to ensure that modelers do not lose sight of the end goals, and structures the model in ways that make it more useful to decision makers and for later use. The process serves as a tool to guide the creative process in much the same way that an EA framework is used to create the EA.

The process of creating a model is not a strictly linear process. All modeling is iterative with updates to the model as more information becomes available and new insights are made. As the model is created, it should be continuously tested and re-evaluated to ensure that it is meeting its objectives. Although testing is often only undertaken toward the end of the modeling process, there is no reason why preliminary tests cannot be performed during model development. After the model is evaluated, new hypotheses can be developed, and the process can be iterated again.

The general form of the process outlined in Table 4.1 is adapted from Sterman's (2000, Chapter 3) process for modeling business dynamics, amended to support a hybrid EA-based approach. The arrows shown indicate feedback loops. Sterman's approach is flexible and oriented toward the development of multipart SD models of businesses, making it easy to adapt for use for creation of hybrid models by emphasizing boundary setting steps and adding in steps specific to hybrid modeling, such as simulation method construction and submodel integration.

4.4.2 Executable Multiagency Enterprise Architectures

EAs are developed to guide investment in systems and services and to guide the improvement of business processes (Sowell 2005). The as-is architecture describes the current business processes, activities, services, data resources, and infrastructure. These are a rich source of initial conditions for state-space parameters. Executable architectures add a modeling capability to analyze the interactions and performance of these elements in the systems and the operational (system of systems) environment. A multiagency executable architecture is formulated in a case study for an international trade scenario in Chapter 11.

Table 4.1   Steps of the Hybrid, EA-Based Simulation Modeling

1. Document the EA.
   - Use an EA framework that supports dynamic modeling, defining views and their interactions.
2. Articulate problem.
   - Is this the proper approach to answer this problem?
   - Identify: problem, key variables, critical behaviors, time horizon.
3. Form a dynamic architectural hypothesis.
   - Initial hypothesis generation.
   - Endogenous focus.
4. Identify the applicable architectural views.
   - Down-select EA views based on problem dynamics.
5. Match views with simulation methodologies.
   - Match based on context of dynamics, required inputs/outputs.
   - Select a modeling environment to be used.
6. Identify boundaries and interfaces among view submodels.
   - Use EA to identify boundaries and interfaces relevant to the problem and dynamic hypothesis.
   - Create submodel boundary charts.
7. Create submodel and top-level model diagrams.
   - Create submodel diagrams of the structure of each submodel.
   - Create top-level model diagram that depicts how the submodels will be linked to create the top-level model.
8. Create the simulation model.
   - Estimation of variables, relationships, and initial conditions.
   - Model subviews with selected methodology.
   - Combine submodels into an architecture model.
   - Develop global model interface.
9. Test model.
   - Sensitivity analysis.
   - Behavioral and structural analyses.
   - Other analysis methods.
10. Design and evaluate policy.
    - Scenario analysis.
    - "What if" analysis.

Source: Adapted from J.D. Sterman, *Business Dynamics: Systems Thinking and Modeling for a Complex World*, p. 86, 2000.

Note: Arrows indicate major feedback loops.

Specific questions that may be addressed with executable architectures include:

- How might a planned portfolio of acquisitions improve mission and business performance?
- How should the acquisitions, process changes, and organizational changes be staged and coordinated in a dynamic transformation process to attain the to-be enterprise state, recognizing that the to-be state is a moving target in an environment of continuous change?

4.5 Unifying Approach to Enterprise Dynamics—Future Directions

Just as fluid dynamics and structural dynamics theories and methods support engineering disciplines, enterprise dynamics supports the ESE/A process as a discipline. All of these fields of application employ a variety of modeling methods depending on the specific nature of the challenge. The concepts and specific methods that comprise an emerging field of enterprise dynamics have been outlined earlier. This section describes a concept for integrating models of specific elements of the enterprise into a comprehensive (unified) view to further advance the application of enterprise dynamics to the ESE/A process.

4.5.1 Unifying Concept

The unifying concept is proposed as a future capability for ESE/A to stimulate further research. The concept is based on a state-space description of dynamic elements of the enterprise from a temporal perspective coupled with the C-I-U operational perspective. Its feasibility is demonstrated in the case study in Chapter 8 as a reduced form of an SE process applied to enterprise transformation, reduced form in terms of the number of enterprise state-space and control parameters using mathematical techniques of nonlinear closed-form optimal control theory and game theory.

The structure of the unifying framework to implement this concept captures the transformation path of the enterprise in mathematical state-space form and is illustrated in Figure 4.5, the unifying framework. The representative analytical methods that may be applied are shown as source modules and may be selected as applicable to the enterprise challenge. The unifying framework also includes the core C-I-U enterprise model to analyze the transformation of the enterprise.

The dynamic parameters are drawn from a diverse set of multidisciplinary source modules formulated to address specific acquisition and operational issues as needed for a specific analysis. Source modules are defined here as models based

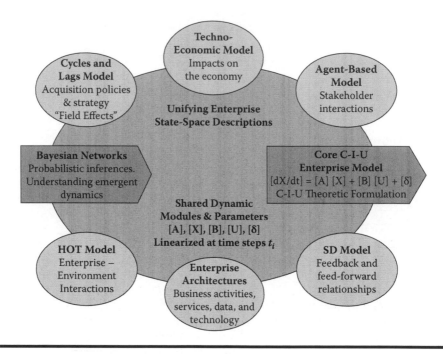

Figure 4.5 The unifying framework includes source modules that contribute to the state-space description of the enterprise, integrates these into a complete state-space description, and uses the core C-I-U enterprise model to analyze the transformation.

on specific disciplines or methodologies that are applied to selected elements of the enterprise. The dynamic parameters in these modules can be expressed as point estimates, statistical parameters, a time series, or larger modules.

The C-I-U enterprise model is formulated here as a piecewise linear control-theoretic problem of the form:

$$\frac{dX}{dt} = A \cdot X + B \cdot U + \delta$$

where
  X = State-space vector description ($n \times 1$)
  A = State-space coefficients ($n \times n$)
  U = Control elements ($m \times 1$)
  B = Control coefficients ($n \times m$)
  δ = Perturbing elements of state-space ($n \times 1$)

Source modules shown in Figure 4.5 can contribute to a comprehensive enterprise state-space description, from which appropriate versions of the core C-I-U enterprise model may be constructed. This C-I-U formulation encompasses all of the controlled-influenced-uncontrolled elements that may be defined by the B matrix coefficients. The control-theoretic approach is necessary to capture the most restrictive case of controlled elements applicable to a specific enterprise.

The piecewise linear formulation is employed to incorporate dynamic state-space parameters from a wide variety of analytical models representing different disciplines and perspectives. Rabelo and Speller (2005) have demonstrated this linearization process for an SD model of enterprise operations. The same approach can be applied to agent-based, discrete-event, econometric, and other types of models to capture their representation of enterprise state-space dynamics. Probabilistic behaviors are important and may also be captured in this framework but further research is needed to improve the efficiency of such methods.

The core unifying model is formulated at the enterprise operations level, incorporating the dynamics of the acquisition program and the influence of the changing external environment. The control-theoretic approach provides powerful interpretive methods involving concepts of observability and controllability as well as the derivation of the eigenvalues of state-space and control matrices.

To provide effective management and decision support, descriptive enterprise dynamics must project the evolution of the enterprise in its state-space and accommodate specific decision-related management control actions. Enterprise state-space descriptors (X) include financial and physical resources of the enterprise and its operating parameters. Control parameters (U) represent management decisions and actions to direct the transformation and operations. Examples of these parameters include

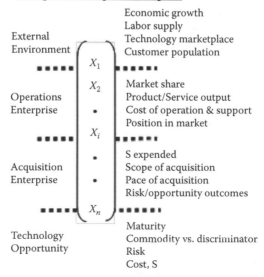

Enterprise State-Space Descriptors

External Environment
- Economic growth
- Labor supply
- Technology marketplace
- Customer population

X_1

X_2

Operations Enterprise
- Market share
- Product/Service output
- Cost of operation & support
- Position in market

X_i

Acquisition Enterprise
- S expended
- Scope of acquisition
- Pace of acquisition
- Risk/opportunity outcomes

X_n

Technology Opportunity
- Maturity
- Commodity vs. discriminator
- Risk
- Cost, S

Enterprise Control/Influence Parameters

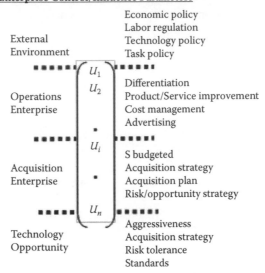

External
Environment

Economic policy
Labor regulation
Technology policy
Task policy

U_1
U_2

Operations
Enterprise

Differentiation
Product/Service improvement
Cost management
Advertising

U_i

Acquisition
Enterprise

S budgeted
Acquisition strategy
Acquisition plan
Risk/opportunity strategy

U_n

Technology
Opportunity

Aggressiveness
Acquisition strategy
Risk tolerance
Standards

Many of these state-space descriptors and control parameters can be drawn from the features of the EA, or conversely, should be introduced into the architecture. This provides the direct coupling of enterprise dynamics to management methods such as EA, lean Six-Sigma, enterprise resource planning (ERP), and activity-based management, and advances the ESE/A process.

Individual source modules (models) define selected descriptors of interest in the enterprise state-space as described above. Their value is enhanced by "unifying" their respective contributions to the state-space characterization. Unifying the various model parameters may involve a comparative selection process when different models cover common elements of the state-space or in an integrative process when the models provide complementary coverage.

Work to date, as described in Chapter 8, has addressed the formulation of a highly aggregated C-I-U model of enterprise dynamics employing elements of the HOT and cycles and lags methods in a game-theoretic representation of the enterprise evolving through a major acquisition in its dynamic external environment. The piecewise linearization of the HOT model is described, and provides a linkage to the greater state-space detail developed in other (source) models.

As source modules are fully developed and validated for a specific analytical purpose, they may be incorporated in a more complete enterprise state-space description (the C-I-U model) of the enterprise. The unifying process using the piecewise linearization of diverse models (Rabelo and Speller 2005) will help in navigating across the scale levels and through the full temporal life cycle to implementation in the operational mission environment of the enterprise.

4.5.1.1 Enterprise State-Space Description and Relation to Architectures

The state-space descriptions in the C-I-U model are a shared resource of enterprise dynamics parameters derived from and related to the EA and employing various disciplines in the source modules. The parameters (X, U, A, B) are time dependent; they can be described analytically in closed form in some cases, but in more complex situations they could be based on piecewise linearization of the various model parameters at specific time steps that retain nonlinear dynamic effects.

4.5.1.2 The Core C-I-U Enterprise Model

The core C-I-U enterprise model uses the control-theoretic approach that integrates selected source modules to address the following enterprise-level questions:

- What is the expected transition progress to be achieved given critical decisions on scope, strategy, and time pressure?
- What are the effects of controlled, influenced, and uncontrolled factors, and the changing external environment on requirements and the program?

The illustrative result of a simplified formulation draws primarily on selected state-space descriptors from the HOT, cycles and lags, and SD models. It provides a highly aggregated description of the enterprise operating within a changing environment, mapping expected progress during acquisition and implementation of new enterprise capabilities. The core model must be sufficiently flexible to incorporate additional dynamic state-space elements from other source modules and to increase its level of fidelity based on such elements when required to support specific management decisions.

In the generalized concept of the unifying framework, the structure of the core enterprise model will be based on the specific planning and analysis study to be performed. The core model may be constructed around one of the source modules or, as in this illustration, based on a unique modeling approach. In either case, content from other source models will be drawn upon and integrated as needed to provide the more comprehensive state-space description of the enterprise dictated by the application.

4.5.1.3 Relationship of Enterprise Dynamics to Enterprise Architecture

Figure 4.6 illustrates the dynamic elements of the enterprise, each with specific time constants, and cycles and lags. Organizational factors including roles, responsibilities, accountability, and authority (R2A2s) are particularly important for

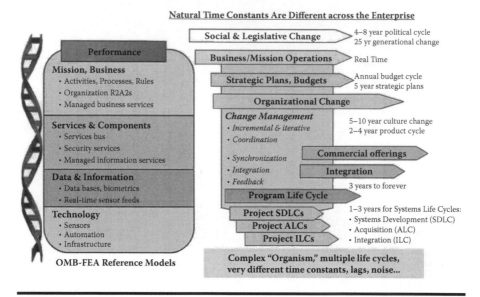

Figure 4.6 Dynamic elements of the EA for an enterprise in transformation.

governance. The dynamics state-space descriptors and control parameters in the unifying approach may be directly related to the dynamic features that must be incorporated in the EA (shown in Figure 4.6 as the DNA of the enterprise).

The EA, as one of the source modules in Figure 4.5, plays an important role as the management tool where services to be delivered and related enterprise activities are established and defined for consistent use across the ESE/A process.

To gain the full benefits of enterprise dynamics coupled with enterprise architecting in the ESE/A process, a consistent set of dynamic state-space descriptors and control parameters must be incorporated. The formulation involves a highly iterative process between dynamic analytics and architecting. The benefits of such coordination address several persistent problems that have been observed in applying these and other management methods independently:

■ Simulation models too often lack content of high importance to the business operations and management.
■ Enterprise architectures contain a rather complete set of enterprise characteristics including business rules (or policies), organizational definition, service or product outputs, activities, and processes, but are generally too static and rigid in their formulation and lack control (and influencing) mechanisms.
■ Activity-based management is generally coupled with a need for an activity-based costing system limiting their use as a general management approach.
■ Enterprise resource planning is often used for financial reporting in a narrow sense rather than as a robust and effective tool for performance monitoring, measurement, and management.

The ESE/A process can resolve these issues, principally by propagating a consistent definition of services and activities throughout the various tools and methods, still not easy in application, but well worth the effort given the very high risks and failure rates observed in even simple transformation programs.

4.5.2 Results of the C-I-U Simulation of the Enterprise in Its Environment

For illustrative purposes, the basic mathematical formulation and results of the C-I-U Model Version 1.0, which was constructed as a control-theoretic representation drawing on selected state-space parameters in the source HOT model and the cycles and lags model, are outlined below. This illustration is drawn from the more extensive case study on this topic in Chapter 8 and demonstrates the feasibility of the C-I-U analytic approach.

In its simplest form applied to the entire scope of enterprise acquisition and use in operations, the core model has a time-dependent scalar state $s(t)$ and controls $q(t)$ and t_f. Here, $s(t)$ quantifies the progress made toward acquiring an operational capability and $q(t)$ is the square root of money spent toward the acquisition. The control parameter t_f is the time to finish the acquisition. In this simple model "acquisition" encompasses transition to operations as well as the traditional acquisition process. Also, a total system lifetime (across both acquisition and operational use) t_L is specified as are constants b and K in the control formulation:

$$\dot{s}(t) = b \cdot q(t)$$

$$J = K \cdot s(t_f) \cdot (t_L - t_f) - \int_0^{t_f} q^2(t) \cdot dt$$

Here, J is the objective function to be maximized. The first term in J represents the benefit of the capability in operations, and the second term represents total cost to acquire the capability. Wojcik and Hoffman (2007) demonstrate that $s(t)$ at optimum is linear in t, and, remarkably, that the optimum t_f is always $t_L/3$, independent of the values of the constants K and b. This very general rule provides a quantitative benchmark against which all kinds of enterprise acquisition programs can be compared. Wojcik and Hoffman (2007) also extend the core model to encompass exogenous effects due to environmental changes, evolving user requirements, or an unstable enterprise mission. Thus, through the application of optimal control theory, the core model generates high-level, emergent enterprise dynamic behavior in the context of its mission and its environment.

4.6 Conclusion

This chapter presents a series of models that can be drawn upon for ESE and positions them to support dynamic enterprise architecting through the ESE/A process. Although this establishes an improved capability for planning and managing enterprise transformation, a possible future direction—the unifying concept—is described to motivate further research. Individual modeling methods tend to be used for specific purposes in enterprise dynamics and are shown with respect to the POET space in Figure 4.7. The positioning indicated represents the more common application tendencies for each modeling approach but many excellent examples of each can be found that have a wider range of application. Such specialized tendencies account for the popularity of hybrid modeling in the Section II case studies and emphasize the need for integration (unifying) methods.

The operational performance benefits from the introduction of enterprise dynamics are an essential part of enterprise planning and analysis for the acquisition and implementation of technology (i.e., in support of the ESE/A process). The proposed framework describes the scope and content of this emerging engineering discipline.

The application of performance simulation models with enterprise architecting described in this chapter is a state-of-the-art ESE/A process. This method can be advanced further via the unifying concept as that framework and the core C-I-U model matures.

The definition of the core C-I-U model at the enterprise operational scale and the definition of dynamic sense-and-respond mechanisms provide unifying capabilities directed both at the higher scale of the economic sector in which the enterprise is embedded and the lower organizational scales. Fidelity may be added to the core C-I-U model as required for analysis as the source modules are fully developed to provide more detailed and complete state-space descriptions.

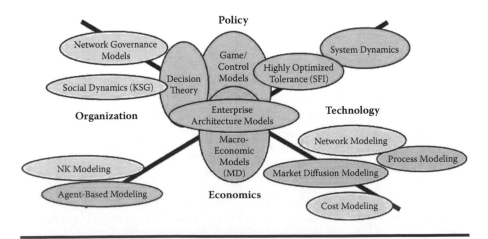

Figure 4.7 Models mapped into the POET space to reflect tendencies in applications.

Methods of tagging the state-space descriptors to identify their temporal position in the life cycle, modular size, scale, and other perspectives should be investigated further, for example, via the use of semantic technology. Improved methods to address uncertainties and to incorporate probabilistic factors in the framework are needed for decision support and should be coupled with visualization and gaming capabilities.

Several of the methods represented in the source modules address dynamic interactions of the enterprise with its external, mostly uncontrolled, environment, such as systems dynamics, HOT, and agent-based models. Future research will explore modules describing competitive, adversarial, and deceptive behaviors as important dynamic elements.

The models discussed earlier have been validated based on extensive and diverse programmatic and operational management experience within the research program. Validation must be extended to the agency planning and program management communities. This may be accomplished through an acquisition gaming environment that brings stakeholders and decision makers together to explore a situation and solutions supported by a diverse set of descriptive, prescriptive, and predictive models with a core state-space representation.

References

Argyris, C. and Schön, D. 1978. *Organizational Learning: A Theory of Action Perspective.* Reading, MA: Addison-Wesley.

Bousso, R, and Polchinski, J. 2004. The string theory landscape. *Scientific American* (September): 78–87.

Bozdogan, K. 2007. Enterprise architecture modeling, design and transformation: Defining the missing links, *Lean Aerospace Initiative*, April, Cambridge, MA: MIT.

Bozdogan, K., Glazner, C., Hoffman, K., and Sussman, J. 2008. Computational Enterprise Modeling and Simulation, Lean Advancement Institute Annual Conference, Boston, MA (April 23).

Bunting, W.J. 2012. Reasoning on uncertain enterprise technology alignment for insight into attainment of enterprise transformation, *Journal of Enterprise Transformation*, 2(1, March): 50–79.

Carlson, J.M. and Doyle, J. 1999. Highly optimized tolerance: A mechanism for power laws in designed systems. *Physical Review E*, 60(2, August): 1412–1427.

Crutchfield, J.P. 2003. When evolution is revolution—Origins of innovation. In J. P. Crutchfield and P. Schuster (Eds.), *Evolutionary Dynamics: Exploring the Interplay of Selection, Accident, Neutrality, and Function.* Oxford: Oxford University Press.

Dahlgren, J. 2007. Real options and flexibility in organizational design. Presented at the Fifth Annual Conference on Systems Engineering Research (CSER), International Council on Systems Engineering (INCOSE), Hoboken, NJ (March 14–16).

DeRosa, J.K. 2005. Thoughts on complex systems and enterprise systems engineering (ESE). Presentation to the Faculty of the College of Engineering and Mathematical Sciences at University of Vermont, Burlington. December.

Epstein, J. 2003. *Growing Adaptive Organizations: An Agent Based Computational Approach.* The Brookings Institution. Washington, DC.

Ethiraj, S.K. and Levinthal, D. 2004. Modularity and innovation in complex systems. *Management Science,* 50(2): 159–173.

Flint, D. 2005. The users view of why IT projects fail. Gartner, Inc. Research Paper (4 February). Washington, DC.

Fowler, A. 2003. Systems modelling, simulation, and the dynamics of strategy. *Journal of Business Research,* 56(2): 135–144.

Galbraith, J. 1973. *Designing Complex Organizations.* Reading, MA: Addison-Wesley.

Hoffman, K.C. et al. 2007. Descriptive enterprise dynamics—A multi-disciplinary unifying framework. Presented at the Fifth Annual Conference on Systems Engineering Research (CSER), International Council on Systems Engineering (INCOSE), Hoboken, NJ.

Jorgenson, D. 1998. Economic and technological models for evaluation of energy policy. In *Growth,* Vol. I. Cambridge, MA: MIT Press, Chapter 9,

Kalpic, B. and Bernus, P. 2002. Business process modelling in industry—The powerful tool in enterprise management. *Computers in Industry,* 47(3, March): 299–318.

Laskey, K.B. 2008. MEBN: A language for first-order Bayesian knowledge bases. *Artificial Intelligence,* 172(2–3): 140–178.

Levinthal, D.A. 2001. Dynamics of organizations. In *Modeling Adaptation on Rugged Landscapes.* Cambridge, MA: MIT Press, Chapter 11.

Love, G. 2006. emRAM—Enterprise modernization assessment model. Briefing presented at MITRE, McLean, VA (20 October).

Mathieu, J., Mahoney, P., and White, B. 2006. Agent-based acquisition stakeholder model. Briefing presented at MITRE, Bedford, MA.

Rabelo, L. and Speller, T.H., Jr. 2005. Sustaining growth in the modern enterprise: A case study. *Journal of Engineering and Technology Management,* 22: 274–290.

Rahimzadegan, B., West, H., and Hoffman, K.C. 2006. HRW cycles and lags model. Vers. 2.0. Briefing presented at Cambridge, MA.

Rivkin, J.W. and Sigglekow, N. 2003. Balancing search and stability: Interdependencies among elements of organizational design. *Management Science,* 49(3): 290–311.

Ross, J.W. et al. 2006. *Enterprise Architecture as Strategy.* Cambridge, MA:Harvard Business School Press.

Schieritz, N. and Größler, A. 2003. Emergent structures in supply chains: A study integrating agent-based and system dynamics modeling. *IEEE Computer Society: Proceedings of the 36th Hawaii International Conference on System Science.*

Schum, D.A. 1994. *Evidential Foundations of Probabilistic Reasoning.* Hoboken, NJ: John Wiley & Sons.

Senge, P.M. 1990. *The Fifth Discipline: The Art and Practice of the Learning Organization.* New York: Doubleday.

Sigglekow, N. and Levinthal, D.A. 2003. Temporarily divide to conquer: Centralized, decentralized, and reintegrated organizational approaches to exploration and adaptation. *Organization Science,* 14: 650–669.

Snowden, D. and Stanbridge, P. 2005. The landscape of management: Creating the context for understanding social complexity. *E:CO* 6(1–2): 140–148.

Sowell, P.K. 2005. A readiness model for multi-agency interaction. MITRE Technical Report, McLean, VA.

Sterman, J.D. 2000. *Business Dynamics: Systems Thinking and Modeling for a Complex World.* Boston: Irwin McGraw-Hill, p. 137.

Stevens, R. 2010. *Engineering Mega-Systems: The Challenge of Systems Engineering in the Information Age*. Boca Raton, FL: CRC Press.

White, B. 2006. On the pursuit of enterprise opportunities by systems engineering organizations. Presented at the IEEE Conference on Systems of Systems Engineering, Los Angeles (April 24–26).

Wojcik., L.A. 2004. A highly-optimized tolerance (HOT) model of the large-scale systems engineering process. In *Student Papers: Complex Systems Summer School* (June 6–July 2). Santa Fe, NM: Santa Fe Institute.

Wojcik, L.A. and Hoffman, K.C. 2007. Emergent enterprise dynamics in optimal control models of the system of systems engineering process. Presented at the 2007 IEEE/SMC International Conference on System of Systems Engineering (SoSE) (April 16–18).

Yates, F.E., MD, 2002. From Homeostasis to Homeodynamics—Energy, Action, Stability, and Senescense, gyates@ix.netcom.com. (Synopsis; November 11).

Chapter 5

Managing Enterprise Transformation Using ESE/A

Fran Dougherty, Elaine S. Ward,
and Kenneth C. Hoffman

Contents

5.1 Introduction

The management of enterprise operations and transformation—with clear roles, responsibilities, authorities, and accountabilities (R2A2) for performance—is central to successful outcomes. Quantitative and qualitative analytical tools such as those described in earlier chapters are often too isolated from management processes. In order to take advantage of the principles of enterprise dynamics and the enterprise systems engineering and architecting (ESE/A) process to improve performance, they must be more closely related to current mainline enterprise management methods such as activity-based management (ABM), enterprise architectures (EA), and enterprise resource planning (ERP).

This chapter on the management of enterprise transformation emphasizes the "A" for architecting in the ESE/A process. EAs are in common use in the public and private sector and, in successful applications, involve participation across the enterprise. Thus, architectures provide an existing and effective linkage to governance and management that can be strengthened. To illustrate a challenging enterprise environment that requires a high level of management coordination, the presentation draws upon an application to the international trade supply chain that is described more completely in Chapter 11. This case study describes a challenging multiagency enterprise environment that requires a high level of management coordination and a clear governance structure.

Architecture-based methods can be enhanced to support effective management of a public or private enterprise by drawing on dynamic models of enterprise performance as outlined in the overall ESE/A process. The paradigm for this application is a sense-and-respond decision-oriented organization with specific attention to elements of the enterprise that can be controlled, internal and external elements that can be influenced, and some internal but largely external uncontrollable factors in the larger environment surrounding the enterprise that must be considered, anticipated, and dealt with.

Analytical and judgmental elements must be embedded in the ESE/A process for application across the enterprise through the architecting process. These elements help to coordinate planning and management better across the multiple organizations and agencies involved in complex undertakings. Security missions of national importance and the delivery of comprehensive public services require increased integration of multiple government, public, and private sector activities. Global supply chains are formed and operated by commercial entities. These collaborative environments pose complex policy, organizational, economic, and technology challenges that require effective management methods and technologies for improved mission performance and to manage risks: a governance challenge!

A systems architecting perspective on governance can ensure that the critical elements of an enterprise transformation program are aligned with operational mission objectives and the use of a distributed resource base. The multiagent business model is increasing in importance in both public and private sectors and poses the most challenging governance requirements; it is used as a reference point throughout this chapter.

5.2 New Directions in Managing Enterprise Transformation Involving Complex Systems

Earlier chapters described a structured approach to architecting, engineering, and managing enterprise transformation programs. The approach must support and respond to changes in management perspectives and approaches when dealing with complex systems and transformation.

In Chapter 2 we presented an historical overview of the conceptual and methodological timeline leading up to ESE/A. We then expanded this discussion in Chapter 3, where we explored the interplay between ESE/A and the emerging system-of-systems (SoS) thinking and the merging with organization science concepts, and spelled out its ramifications for the discipline of ESE/A, with special emphasis on the application of ESE/A for designing and managing changes in large-scale complex sociotechnical systems. Then, Chapter 4 outlined a variety of modeling methods drawn from operations research, management science, and economics to explore the dynamics of enterprise transformation and to deal with inherent uncertainties. Having laid down the essential fundamental ideas in these previous chapters, we can now address the "how-to" question by focusing directly on the ESE/A process (to explore and define structured approaches to managing enterprise transformation) the task of actually implementing ESE/A as a managed process in order to design, engineer, and bring about enterprise performance improvements, modernization, and transformation.

There is no commonly accepted process or structured approach to implementing approaches such as ESE/A. Unlike the progression of structured approaches employed over time to implement systems engineering, a new process for implementing ESE/A must recognize enterprises as goal-directed dynamic adaptive sociotechnical systems that are typically characterized by nonlinear interactions and causal loops, self-organization, and emergent behaviors. Hence, the process-related ideas proposed in this chapter address the challenge of planning the design process in an age of persistent and rapid change, thinking about and managing a heavily interactive and iterative process efficiently and effectively while allowing for continuous organizational learning and adaptation. Although the theme of this chapter is the management of enterprise transformation, it is written from the perspective that transformation is an ongoing process that must be managed simultaneously with uninterrupted business operations.

Therefore, in this chapter we start with earlier systems engineering processes that have evolved over the years from linear or sequential processes to waterfall processes, then to concurrent engineering processes that have morphed into a spiral process, as the very process itself has had to deal with managing steadily increasing system hardware and software complexity. This has meant the incorporation of a growing number of feedback loops to account for the steady growth in system complexity, and, hence, the adoption of an iterative dynamic approach to accomplishing the underlying design process. This evolutionary development in the systems engineering process is reflected in the progressive changes over time in the defense acquisition process and, consequently, in the design and development of complex new products. Both the systems engineering process and the product development process have also been influenced to a significant degree by the evolving processes employed in the design and development of software systems.

These developments pose some important questions. What does it mean to engineer or architect complex enterprises, which represent a subset of complex sociotechnical systems that often exhibit dynamic network characteristics, nonlinear interactions, emergence, and self-organization properties? How does one design and develop, simultaneously, complex engineered systems and the complex sociotechnical systems embodying them with the assurance that the engineered systems will be error-tolerant, particularly in the face of the orders of magnitude greater complexity arising from the interactions of the engineered systems with the multiple dimensions and behavioral factors characterizing the enterprise as a whole?

As we have already explored, the answers to these questions require a new synthesis in contemporary thinking and practice. This does not mean shedding the need for certitude, *ex ante* predictability of how the engineered systems will perform and behave. In today's complex interdependent environment there is all the more reason to expect such complex engineered systems to be fault-tolerant and perform their missions with efficiency, consistency, and reliability.

One strong implication of such a view is that designing and developing complex engineered systems per se should not be undertaken in isolation by considering the total enterprise as a "wraparound" surrounding the outer shell of the engineered system being developed with the hope that the engineered system's interactions with the rest of the enterprise will somehow prove robust to various types of systemic failure. This might be called a *passive design strategy*. Rather, it can be argued that a deliberate effort must be made to endogenize or internalize key aspects of the larger enterprise in the very design process itself, by adopting a proactive "enterprise-in-the-loop" design strategy. This might be called a *proactive design strategy*.

The traditional idea of an all-knowing "system engineer" or "system architect" driving the process, acting on behalf of the customer, end-user, or builder, has given way to the idea of an "open design" process that continuously accepts multiple inputs from all stakeholders who have a clear vested interest in the outcome, through two-way feedback–loop interactions.

The greater complexity inherent in the design-build process indicates the need to bring more science to the design and engineering process itself. We have already seen a significant shift in recent years toward greater utilization of computational modeling and simulation methods in conjunction with system design and engineering processes. Mapping out a robust future process for implementing ESE/A might consider the role and application of multiagent-based modeling, evolutionary computation, optimal control, and Bayesian networks.

There is also a need to bring more science to the test engineering process to uncover and understand the complexity of the sociotechnical systems embodying the engineered systems. This means integrating knowledge related to the structure and dynamics of complex sociotechnical systems (e.g., complexity theory, network science, computational organizational science, evolutionary biology, behavioral economics, etc.). In this context EA—not the architecture of information systems and technologies but the enterprise's architecture as a holistic dynamic system—is gaining increasing ground.

A holistic integrated framework for the ESE/A implementation process can be a unified structured approach for designing and managing enterprise change and modernization and transformation through real-time consideration of policy, organizational, economic, and technological (POET) dimensions of operating complex large-scale enterprises with emergent properties, changing landscapes, and uncertain futures.

A useful starting point for discussion of the ESE/A process is the perspective of the Office of Management and Budget (OMB) on performance planning and budgeting for large-scale federal investments in complex systems and technology

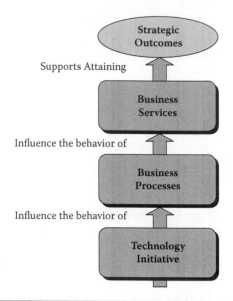

Figure 5.1 **Performance-based line-of-sight from technology initiative through transformed business processes and services to achieve strategic transformation outcomes. (From W. J. Bunting,** *Journal of Enterprise Transformation,* **2(1, January): 50–79, 2012. With permission.)**

initiatives across civil and government agencies as indicated in Figure 5.1 (Bunting 2012). Heavily streamlined approaches following similar logic are applied in the commercial sector.

5.3 Management Challenges[*]

Government and commercial operations increasingly require the integration of a complex mix of business activities and services from support organizations to perform missions (and deliver services) of high national importance. When these operations are modernized by applying advanced technology and information systems, there are unique governance challenges that demand new management capabilities. Again, enterprises are generally in a state of continuous change during which operations must be sustained.

In transformation programs, technical managers must deal with the selection, acquisition, introduction, and operation of these transforming technologies and

---

[*] © K. C. Hoffman et al. 2005. International Enterprise Distributed Object Computing Conference (EDOC) Workshop. IEEE Computer Society.

the associated changes in business practices across all involved agencies. Prominent examples of multiagency missions and the delivery of critical services include:

■ Providing supply chain management for imports and exports by commercial manufacturing and distribution partnerships
■ Managing security aspects of international commerce (Departments of Homeland Security, State, Commerce, and commercial supply chain partners), which is used as a case study
■ Delivering healthcare and services to the elderly and low-income population (Departments of Health and Human Services, Veterans Administration, healthcare providers, and insurance industry)
■ Conducting joint military missions (Department of Defense and logistic support organizations)
■ Providing e-government services to the citizen (all federal and state/local agencies providing citizen services)

The emergence of the collaborative nonhierarchical work environment described by Malone (2004) provides a more general management example of coordinating highly distributed activities. Supporting technologies can provide the required accountability and sharing of information as communities of interest coalesce around a specific mission or business challenge.

A number of enterprise management tools are available and, with appropriate content and application, can help managers deal with the system complexities of planning and transformation for multiagency missions. These tools include EAs, simulation models, and ERP methods. The architecture-based approach to management in the multiagency environment can integrate these methods, and is developed in the following sections that first address:

■ The management domain for transformation (5.4)
■ Governance in the multiagency environment which is of increasing importance as agencies and business must interact to perform complex mission functions (5.5)
■ Steps to build an effective governance environment to manage complex transformations (5.6).

These lead to ESE/A processes that support the full life cycle of enterprise transformation including:

■ Background on ESE/A process guidelines drawn from MITRE and MIT experience (5.7)
■ A holistic ESE/A process employing enterprise dynamics tools and methods (5.8)

5.4 The Enterprise Management Domain— Transformation Methods and Experience

Any sufficiently advanced technology is indistinguishable from magic.

Arthur C. Clarke

Enterprise transformation involves the implementation of new technologies and business processes to improve performance of the mission and support activities of the enterprise. Transforming technologies include automation and sensor technologies applied directly to performance of the mission or delivery of a service and information technologies that provide the mission or service with better decision support, data, and communication systems.

The states of the planning and investment processes for selecting and deploying advanced technology are mature for individual business entities, but the failure rates remain unacceptably high. Federal, state, and local governments and private sector enterprises are investing in EAs to better manage transformation programs. The architecture may be viewed as the genomic code of the enterprise and describes the mission, organization, and business processes along with the supporting information systems and infrastructure.

Activity-based costing (ABC) and ERP methods are commonly used to consolidate large-scale financial and accounting systems that track enterprise performance and the progress of transformation initiatives. ERP systems are used extensively today, however, they are not customarily coupled with the architectural view of the enterprise. In the governance environment described in this chapter, ERP systems can be used to capture an always-current state-space description of the enterprise. Collectively, these ABC and ERP methods and practices provide a foundation for agency-level planning and transformation but must be extended, integrated, and adapted to the more complex multiagency environment. The discipline of enterprise dynamics provides a unifying theory for more effective application of these planning and management methods.

Failures of transformation programs, whether for individual or multiple entities, often stem from governance defects involving such risk factors as:

- Poor coordination among agencies involved in a mission partnership
- Poorly defined mission objectives
- Differing terminology, timeframes, and cultures that hamper and degrade communication
- "Stovepiping" of business, financial, performance, and information system planning and management, where separation leads to differing definitions of critical business activities

- Performance objectives that are not related directly to business/mission effectiveness or quality and not easily measured
- Lack of vision and specificity in the desired future business environment and the target architecture that will support it, both of which should drive the transformation programs
- Lack of executive and organizational commitment to an "architected" approach despite the large investments in architectures

These issues are much more likely to occur because of organizational structure and the newness of the collaborative experience. For example, management structures may be less clear in multiagency operations or when there is little history of staff working together. With technical systems and data sources distributed across organizational units and locations, the situation lends itself to a high risk of failure.

An integrated governance approach based on architectural concepts can help to avoid some of these pitfalls by providing a complete structure for planning and decision making. If applied with full management commitment, the ESE/A process will also help to make the architectures more relevant and manageable for implementation. The role of the technical manager is particularly important in developing and directing plans for transformation. The complexity of the environment requires that the technical managers make good use of every effective enterprise tool at their disposal.

A variety of enterprise planning and analysis resources are employed to support governance processes for both mission operations and transformation programs including:

- *EAs*, describing all tangible, and some intangible, aspects of the enterprise
- *Operations models and simulations,* describing the performance of business and mission organizations supported by information services
- *ERP and ABM methods/processes*, encompassing all enterprise resources applied to the mission (human, technology, facilities, etc.) for performance monitoring

In a single agency or enterprise, the governance of enterprise transformation is based on its organizational structure and the delegation of authority within clear lines of control. In the multiagency environment where these mechanisms are absent, governance is particularly challenging because it must encompass a set of mission activities and support services that are distributed across agencies and commercial partners, each with its own management processes and culture and no established lines of control.

The challenge of enterprise transformation in the more complex multiagency environment is amenable to an architecture-based approach which can suggest the needed structure of relationships at the business, support service, data, and infrastructure levels.

5.5 Effective Governance in a Multiagency Environment

The importance and long reach of public and private sector enterprises in delivering products and services at a national and global scale motivate our interest in the multiagency enterprise. These involve extensive partnerships and supply chain relationships that add significantly to opportunities and risks and to technical and management challenges. Thus, the multiagency model is becoming a dominant mode with a greater need for the methods of enterprise dynamics. For these reasons, the multiagency challenge is emphasized here as an exemplar application. The responsibilities of managers in a multiagency transformation program must be carried out across participating agencies in close coordination with operations managers responsible for mission performance. The scope of this overall governance framework and the requirements to be addressed by the manager(s) are described below.

5.5.1 The Scope of Governance in Multiple Agencies

Governance, in its largest sense, deals with all management aspects of an enterprise, including stakeholder relations and the public good: "Good corporate governance should provide proper incentives for the board and management to pursue objectives that are in the interests of the company and its shareholders and should facilitate effective monitoring. The presence of an effective corporate governance system, within an individual company and across an economy as a whole, helps to provide a degree of confidence that is necessary for the proper functioning of a market economy. As a result, the cost of capital is lower and firms are encouraged to use resources more efficiently, thereby underpinning growth" (OECD 2004, p. 11).

These principles apply equally to the government sector. The working definition of governance used in this chapter focuses on the important mission-oriented aspects of managing operational activities and transformation programs with significant technological content in a way that meets management's obligations to stakeholders and the public: in short, the efficient use of capital, both human and financial. The following working definition is drawn from the Information Systems Audit and Control Foundation: "The set of responsibilities and practices exercised by the board and executive management with the goal of providing strategic direction, ensuring that objectives are achieved, ascertaining that risks are managed appropriately, and verifying that the organizations resources are used responsibly" (Hamaker 2003, website).

This definition of governance can be applied to the benefit of stakeholders in both the public and private sectors. A sound governance structure is essential in the multiagency environment where responsibility, authority, and accountability for both technical management and operations management are widely distributed. This situation can involve varying degrees of centralization, ranging from an overall program manager (or coordinator) to the other extreme of a distributed collaborative management team. The architecture-based approach applies across this spectrum; it is particularly powerful as a coordinating mechanism in all situations involving a high degree of distributed responsibilities.

5.5.2 Requirements for Effective Governance

An earlier presidential directive (Executive Office of the President 2002) set high-level governance objectives that drove the evolution of government policies, architectures, and financial management systems to better support the government-wide initiatives of:

1. Strategic management of human capital
2. Competitive sourcing
3. Improved financial performance
4. Expanded electronic government
5. Budget and performance integration

Governance of multiagency missions and services requires new concepts and a broad systems perspective to achieve these objectives as well as more consistent planning and information to deal with the close coordination of organizational plans, work (business) processes, and resource management functions as illustrated in Figure 5.2. These functions must be performed by technical and operations managers over the full range of timeframes ranging from the near-term management of performance with existing assets and resources to longer-term improvements through technology modernization. The architecture is shown here in a central role to provide an integrated view of mission activities, services, data, and technology and to indicate how these features of the architecture relate to organizations, resources, and work processes. The interactions of these elements of the dynamic enterprise can be modeled using the methods outlined earlier; specific selections depend on the position of the enterprise in the operational landscape that describes the nature of the mission and applicable decision processes.

Because of the complexity of multiagency interactions and the rich set of emerging technology opportunities, effective governance requires a close coupling of operations management with technical management. Specific requirements of effective governance include the following.

5.5.2.1 Virtual Enterprise Model That Captures Organizational Interactions and Interoperable Systems

Interactions (across organizations, activities, business processes) and interoperability (across data sources, information systems, and infrastructure) are critical to the effective performance of multiagency missions and business services. Formal methods such as the levels of information systems interoperability (LISI) model (MITRE 1998) are in common use as evaluation tools for data and infrastructure interoperability.

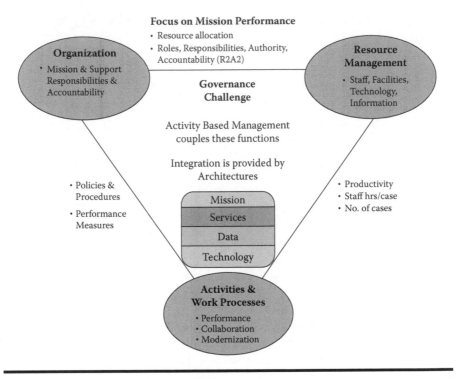

Figure 5.2 The governance challenge: coupling organization management, resource management, and work processes for mission performance.

Clark and Jones (1999) have proposed a five-level model for defining and evaluating interactions at the organizational and activity levels:

■ *Level 5—Unified:* Organizations are completely integrated in day-to-day working and use an integrated knowledge base with a high level of commitment to shared goals.

■ *Level 4—Integrated:* Shared value systems and shared goals are in place for organizational units with a common understanding and preparedness for joint operations in mission performance.

■ *Level 3—Collaborative:* Recognized relationships are in place and shared goals are recognized, but the organizations are still distinct in performing mission activities.

■ *Level 2—Cooperative (Ad Hoc):* Limited organizational interaction on an ad hoc basis with liaison points established for situational awareness of the presence of other agencies.

■ *Level 1—Independent:* Separate organizational structures with little interaction and no information exchange or shared purpose.

Level 5 achieves the goal of a single agency, whereas Level 1 is not representative of multiagency missions. Therefore, the objectives of this chapter apply to Levels 2 through 4. The nature of the multiagency mission and the other responsibilities of the participating organizations should dictate the preferred level. Strong management can achieve the required levels of interaction by using the completeness of architectural representations as a template to guide governance roles and responsibilities. For purposes of this chapter, reference is made to the Integrated level of interaction (Level 4) with the understanding that adjustments can be made to other levels as required, whether for the long term or for shorter periods where another level of interaction might be appropriate.

Modernized information systems enable the virtual enterprise model where roles and responsibilities are distributed across organizational lines and also geographically dispersed but remain tightly integrated through communication channels to a central management entity. Effective interaction and interoperability are provided by an "architected" system for communication and management with strong accountability for assigned activities and the delivery of support services from distributed locations and organizations. These capabilities are important to the success of the manager.

To achieve high levels of mission performance, effective governance requires clear roles, responsibilities, authorities, and accountability for each agency unit involved in the business service or mission. A completely defined set of individual activities performed according to defined processes and prescribed outcomes provides this clarity of purpose and is indeed a central element of an EA for both single and multiagency missions.

5.5.2.2 Well-Defined Business Functions and Processes

Operational management and transformation must address the specific mission/business processes and the specific activities performed by participating agencies, public and private. An integrated multiagency mission and business process view is needed to support all principal executives and staff including:

- The executive manager role of assigning responsibilities and accountability to organizational elements
- The operations manager role of using activity-based performance methods
- The chief information officer role of planning information technology (IT) modernization programs
- The chief financial officer role of financial management, budgeting, and performance measurement
- The technical program manager role in designing and implementing enterprise transformation programs

5.5.2.3 Heterogeneous Technical Services Environment

Agencies will select specific standards, hardware, software, tools, and methods based on their own requirements. Drawing on these elements, the multiagency mission will be carried out in a heterogeneous organizational and services environment for architectures, financial management, and system development.

Specific IT services must be defined and related explicitly to mission activities. The managed service approach, with well-defined operating level agreements (OLA) and service level agreements (SLA) for the selected service provider, supports the selection and integration of diverse agencies and commercial partners.

5.5.2.4 Comprehensive Planning and Investment Methods

In response to the Clinger–Cohen Act (U.S. Congress 2000) and OMB guidelines (OMB 1996; OMB FEA 2003) for technology budgeting, federal agencies are making massive investments in EAs at the agency level and for major bureaus and offices within the agencies. Commercial enterprises, as well as state and local governments, are also developing information system architectures (NASCIO 2002). Although these products are applied widely for IT investment planning and budgeting, they are rarely used as an overall enterprise-level strategic planning and management tool. As such, architectures represent a seriously underutilized resource.

At the same time many agencies are implementing ERP systems to consolidate financial and accounting systems. This capability is important in allocating resources to transformation programs and measuring their operational effectiveness. Although used extensively in this domain, ERP systems lack the type of enterprise data and information that is contained in the architecture and related to mission performance.

A variety of architecture frameworks and approaches has been applied successfully in the commercial and government environments to plan technology investments (Hoffman and Melancon 1988; Zachman 1987). Frameworks in the government sector driven by the Clinger–Cohen Act include the Federal Enterprise Architecture Framework (FEAF) (CIO 2001), and Command, Control, Communications, Computers, Intelligence, Surveillance, and Reconnaissance (C4ISR)/DoD Architecture Framework (DoD 1997). The reference models (OMB FEA 2007) are a more recent development. The performance reference model is the most important addition and motivates the need for dynamic architectures. There are also numerous technical standards concerning computing and communications that are embedded in the technical level of these architectures.

Although many of these early policies and frameworks have evolved, their implementation has remained static and deterministic. The ESE/A process provides a major step forward toward dynamic methods more suitable to the complexities of operations transformation and management.

5.5.2.5 EAs Must Be Compatible across Agencies Participating in Multiagency Missions

EAs for public and private sector entities are a major source of business and technical detail needed for integrated planning of the multiagency mission. Figure 5.3

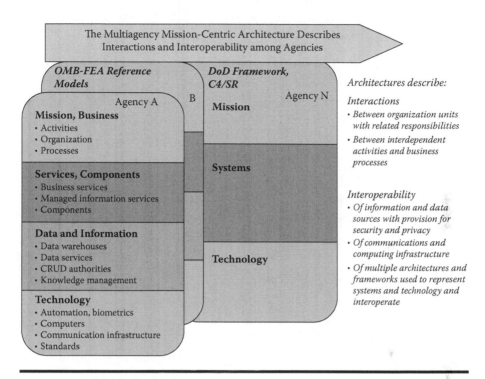

Figure 5.3 Multiagency architecture templates with an integrated perspective.

describes the content of architectures and portrays the required integrated perspective to accommodate individual agency architectures that may use different frameworks. A mission-centric architecture highlights the coordination, cooperation, and collaboration required at all levels of the architectures. Each agency's EA will provide a full enterprisewide information base regardless of the frameworks used, however, there must be a high level of consistency in several key areas to ensure coordination and interoperability among the specific activities performed by each agency in the multiagency mission.

The multiagency governance perspective must integrate EAs of participating agencies with different formats emphasizing the dynamics of multilevel interaction and interoperability. This integration supports a "virtual environment" concept for multiagency missions.

Accomplishing a high level of consistency in these areas requires a complete and consistent set of features at each level of the EA, drawing on enterprise dynamics methods, namely:

■ *Business*: Activities are completely defined in terms of outcomes, resources required, processes, and responsibility assigned to an agency organizational unit. Business or mission performance can be modeled using activity-based process models or other enterprise dynamics methods discussed earlier.

■ *Services and Components:* Services support activities within an agency, and may be shared across two or more participating agencies. The requirements of, or contributions to, mission performance can be described and optimized using enterprise dynamics modeling methods.

■ *Data and Information:* Data support activities within an agency and data shared across two or more participating agencies with specific authorities to create, read, update, and delete data. Critical information needs for decision making and the timeliness of that information can be defined through performance modeling.

■ *Technology/Infrastructure:* Infrastructure supports activities within an agency and may be shared across two or more participating agencies, and may interface with other agencies. The potential role of emerging technologies in improved infrastructure performance is central to the formulation of enterprise dynamics methods.

Interfaces and interoperability must be addressed at all of these levels using the LISI concept discussed earlier. The positioning of architectures as an asset for operational support as well as transformation will also ensure that these assets better represent a performance-based business and mission vision. Such a vision is needed to plan and implement incremental improvements through a longer-term transformation program.

5.5.3 Business Activities Play a Central Role in Governance

The allocation of accountability for management of specific activities performed by individual agencies in a multiagency business or mission is central to effective governance. The tasks of achieving tactical and strategic objectives, assigning responsibility and accountability, managing risks, and making responsible use of resources are best addressed at the activity level. In this context activities represent the work elements that must be performed to deliver a specific business service or product or perform a defined portion of a mission.

Activities, whether performed repeatedly or on a one-time basis, must be defined at a level of specificity that permits their inclusion in a documented and repeatable process using specific staff and technical resources. The individual activities, although distributed across agencies, must be coordinated. Activities can be characterized by documented processes for their execution and include the defined inputs of resources, including staff, facilities, technology, and information. Outputs are also defined for services or work products delivered by the activity.

The precise and consistent definition of activities strengthens the application of ABM, a broad discipline and practice to apply activity information to business performance and resource management (Player and Keys 1999). Activities are a key element because they can be defined with precision, assigned as responsibilities with organizational accountability, managed by traditional methods, and measured for performance. This characterization of activities facilitates simulation modeling of the mission or business process to demonstrate performance and thereby validate the technical and governance approaches.

To attain effective governance, the same clear and consistent definition of activities should be used in operational management, business process development, performance measurement, cost and resource accounting (e.g., ERP), and technology modernization. An activity-based approach can accomplish this, and it fits quite well with proven management processes such as ISO-9000, Six Sigma, and other quality programs.

5.6 Building Effective Governance of Transformation Planning and Management

Building the integrated governance environment for a multiagency mission entails a five-step process. As elements of the transformation program proceed toward implementation and as new circumstances are encountered these steps must be reviewed and updated.

Step 1: Assemble, review, and update agency-level planning resources.

- Identify, assemble, and document pertinent planning resources including agency EAs, strategic plans, performance plans, and supporting information systems.
- Review materials for completeness and consistency in representing elements of the mission.
- Update materials as needed to fill gaps in information and coverage of the mission.

Step 2: Construct and maintain the mission-centric architecture.

- Identify all mission or service activities performed by participating agencies.
- Identify agency activities, their supporting services, data, and technical elements, and select those applicable to the mission operational plan.
- Identify and monitor service providers with defined OLAs and SLAs for the specific agency services to be provided.

Step 3: Assign governance roles, responsibilities, authorities, and accountability.

- Organize and charter a multiagency program organization or office or a virtual office with necessary budget authority.
- Define the level of organizational and technical interoperability required to perform all activities in the mission.
- Analyze and assign mission or business activities to lead organizational elements within and across agencies and clearly defined R2A2s for governance.

■ Assign services, data, and technical (utility services) to organizational elements as service providers within and across agencies with R2A2s defined for governance, supported by OLAs and SLAs defined in the mission-centric architecture.
■ Revise R2A2s as dictated by changes to the mission or performance experience.

Step 4: Conduct mission operations performance analysis and monitor progress.

■ Define operational relationships for the end-to-end mission or service activities including performance objectives and measures.
■ Define technical performance objectives and measures to support end-to-end business performance.
■ Apply appropriate enterprise dynamics modeling methods to project performance improvements.
■ Analyze mission performance with transformation initiatives incorporated individually or as a group using the current situation as a performance baseline.
■ Monitor operational performance as transformation proceeds.
■ Monitor performance of the transformation program using earned value or comparable methods.

Step 5: Develop and update multiagency program plans and policies.

■ Evaluate the organizational and technical aspects of mission performance from the Step 4 analysis.
■ Define improved mechanisms for coordination, cooperation, and collaboration among all parties.
■ Develop or update plans and policies, including transformation plans, and the transition to operations.

The five steps in constructing the governance process can be aided by planning and analyses in coordination with the more comprehensive ESE/A process (described in Section 5.8) and must also be performed in close coordination with agency managers. They must also adhere to enterprise-specific management and regulatory policies. The mechanisms for coordination, cooperation, and collaboration developed in the planning phase should be sustained through future implementation efforts.

A key element in the process is Step 2. The mission-centric architecture (also known and commonly implemented as an EA segment) is the principal planning and management resource. It defines and links mission activities, identifies related activities and supporting systems in agency architectures, and integrates these for use in the mission performance model. It captures the relevant features of agency organizations, activities, and systems that support the mission. It also helps to identify gaps in those agency capabilities.

Developing a governance model that will ensure performance accountability and define relationships between engineering/technical managers and operations managers is described in Step 3. Roles, responsibilities, authorities, and accountability are assigned for all architected elements including mission activities and support systems; these can be allocated between the primary mission agency and supporting agencies when such a distinction exists.

The complexity and risks of multiagency missions can be addressed most effectively through an integrated governance approach where the following elements are present:

- A mission program coordinator, program office, or like entity is chartered with a defined transformation role for a lead technical program manager.
- The multiagency parties have budget authority over elements of agency programs related to the mission activities and support services.
- Close coordination is established between technical program management and operations management for performance improvement.

5.7 Background on Systems Engineering and Architecting Guidelines

This section provides some background on structured systems engineering processes for designing, building, and managing complex technical engineering systems for transformation. These processes have been applied successfully to a number of complex systems engineering challenges. The processes require an effective governance environment as outlined in the previous section. The approaches outlined below are the basis of the holistic ESE/A process described in Section 5.8.

5.7.1 MITRE SEPO Regimen

The Systems Engineering Practice Office (SEPO) at MITRE formulated an approach (Kuras and White 2005)—a "regimen"—to apply to large complex programs. The regimen is supplemented in practice with the POET approach to address the multiple dimensions of complex systems. The basic steps in the regimen are:

1. Analyze and shape the environment.
2. Tailor developmental methods to specific regimes and scales.
3. Identify or define targeted outcome spaces.
4. Establish rewards (and penalties).
5. Judge actual results and allocate rewards.
6. Formulate and apply developmental stimulants.
7. Characterize continuously.
8. Formulate and enforce fitness regulations (policing).

5.7.2 MIT Complex, Large-Scale, Interconnected, Open, Sociotechnical Process

The complex, large-scale, interconnected, open, sociotechnical (CLIOS) process (Sussman 2002) has been used on many complex projects and in undergraduate and graduate level courses at MIT. The process is organized in four major segments with 12 steps in total:

Representation of structure:

1. Describe the system (identify issues and goals).
2. Identify major subsystems.
3. Develop the CLIOS diagram (nesting, layering, expanding).

Behavior:

4. Describe components and links.
5. Seek insight about system behavior.

Evaluation:

6. Refine system goals and identify performance measures.
7. Identify options for system performance improvement.
8. Flag important areas of uncertainty.
9. Evaluate options and select those that behave "best" across uncertainties.

Implementation:

10. Develop strategy for implementation.
11. Identify opportunities for institutional changes and architecture development.
12. Perform postimplementation evaluation and modification.

The enterprise landscape concepts discussed earlier in Chapter 4 support the representation steps of CLIOS by providing multiple management and operational perspectives. The enterprise dynamics modeling methods described in the following sections provide a robust capability for the evaluation steps of CLIOS.

Systems engineering (SE) in the earlier traditional form differs from CLIOS and the rest of the engineering systems curriculum at MIT in that SE is mainly technical, whereas engineering systems includes a broader perspective that encompasses management, social sciences, and engineering, as well as focusing on critical contemporary issues, including climate change, trade issues, world sustainability as a development goal, economics, and equity.

5.8 Holistic Enterprise System's Engineering and Architecting Process

The holistic ESE/A process draws on the SEPO and CLIOS processes to provide an integrated approach for applying the discipline of complex ESE/A in large-scale complex technology-enabled engineering-intensive sociotechnical systems. Both the SEPO and CLIOS processes call for descriptions and characterization of the system or enterprise as a critical step that is emphasized here as an application of enterprise dynamics methods.

The holistic ESE/A process is generalized and must be tailored to address the unique characteristics of enterprise challenges. Ideally, both technical and business managers should participate in the process along with analysts and other specialists. From a management perspective, the process should emphasize the concepts of enterprise architecting and enterprise dynamics presented in this sourcebook.

The process combines and augments elements of the SE methods and structured processes described in Section 5.7 in six major steps. Supporting enterprise dynamics tools and methods are identified with specific steps in applying the ESE/A process, with reference to landscapes and perspectives discussed in Chapter 4 and to modeling methods described in the Section II case studies.

Although the ESE/A process is described in a sequential fashion, in practice the steps are highly iterative and interactive to help technical and business managers converge on perspectives, beliefs, opportunities, plans, and outcome objectives. The process, as in the above examples of other processes, is simplified and generalized for a range of applications. References are included to enterprise plans and materials, tools, and methods described in other chapters and references that can be selected and drawn upon for specific application cases. The six major steps, activities, and supporting analytic tools and methods are described in the following subsections.

5.8.1 Statement of the Complex Enterprise System Challenge and Complexities

The challenge and complexities of enterprise transformation can be stated as follows. Define the scope and major elements of the enterprise and the operational or transformation objectives whether they are to be achieved through a major acquisition or managed change in the complex POET environment.

Reference materials for this descriptive step include organizational plans, business strategies, performance plans, architectures, and transformation plans for the enterprise. Specific activities to be performed for this step include:

■ Identify the scope of operations of the enterprise, the geographic reach, the goods and services provided, and all stakeholders involved in a supply chain for production and delivery.

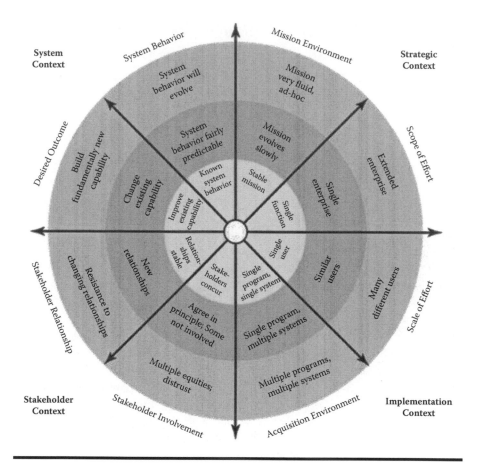

Figure 5.4 (also Figure 4.4)   Enterprise Systems Engineering Profiler.

■ Outline the transformation objectives and elements involved across POET elements of the enterprise.

■ Apply the Enterprise Systems Engineering Profiler, Figure 5.4, to define the transformation complexity in eight dimensions. This activity helps to identify high-risk areas to be managed as the level of complexity increases from the inner ring to the highly complex outer ring descriptions.

5.8.2 Enterprise Characterization and Representation (Architectures, Landscapes)

Following the statement of the transformation challenge and complexities, the ESE/A process calls for a comprehensive representation of structure, resources, and assets of the enterprise and the specific nature of its mission operations including organizational responsibilities, staffing patterns, and process employed, along with policies and procedures that govern mission operations.

Reference materials, tools, and methods for this more detailed descriptive step include operational plans and procedures, enterprise resources and architectures, enterprise landscapes, and other applicable descriptive methods outlined in Chapter 4. Specific activities to be performed for this step include:

- Describe the enterprise structure in process and geographic dimensions.
- Identify roles, responsibilities, and objectives of stakeholders in mission operations and transformation management.
- Describe the elements of the enterprise and interrelationships (identify transition and operations issues and goals).
- Identify major subelements and POET factors: programs, components, and links.
- Develop the ESE/A diagram (a dynamic architecture addressing scale, nesting, layering, expanding of system elements).
- Characterize POET elements of the enterprise.
- Seek insight about POET systems behaviors and interactions.
- Develop an operational perspective that defines controlled, influenced, and uncontrolled elements both internal and external to the enterprise that can affect operations and produce emergent behaviors.
- Apply the fitness landscape to represent temporal dynamics and characteristics of the influenced and external environment to guide sense-and-respond capabilities required for operations and transformation as shown in Figure 5.5.

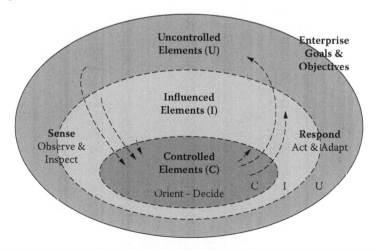

Figure 5.5 (also Figure 4.3) **Operational perspective defining the enterprise in terms of controlled, influenced, and uncontrolled domains, the basis of the core C-I-U enterprise model. (Adapted from J.K. DeRosa, Presentation to the Faculty of the College of Engineering and Mathematical Sciences at the University of Vermont, Burlington. December 2005.)**

5.8.3 Modeling and Analysis Methods (Multidisciplinary Methods)

This analytical step builds on the previous descriptive ESE/A steps to build and apply executable models of enterprise operations in their current and transformed states. This analysis supports planning, analysis, and evaluation of mission performance and cost-effectiveness/benefits analysis of transformed operations. Modeling methods can be coupled with the rich descriptive content of enterprise architectures and ERP systems to provide a decision support tool.

Reference materials, tools, and methods for this more detailed descriptive step include descriptive work products from Step 2, augmented with enterprise dynamics source modules and the unifying approach outlined in Chapter 4. Specific activities to be performed for this step include:

- Refine enterprise system goals and represent as performance metrics and measures to be incorporated in an enterprise operations model.
- Identify the state-space elements of the enterprise including options for enterprise performance improvement drawing on the unifying approach.
- Highlight important areas of uncertainty and risk for inclusion in models drawing on the Profiler (Step 1) and mapping to POET dimensions of the enterprise.
- Apply selected enterprise dynamics source modules to evaluate system behavior and options and select those that behave "best" across uncertainties, considering complexities across the POET elements and the relative strength of specific models across this space as shown in Figure 5.6. (Note that this mapping highlights the relative tendencies of the approaches; there are examples of more comprehensive coverage of POET dimensions than shown for specific models.)

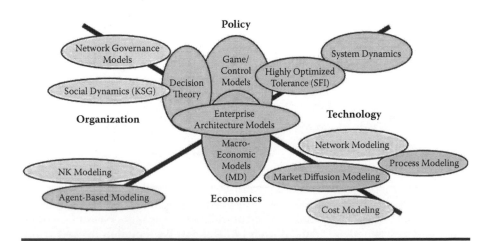

Figure 5.6 (also Figure 4.7)   Mapping of models into POET domains.

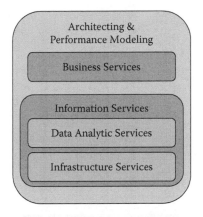

Figure 5.7 Types of services described in an enterprise architecture.

Other tools and methods to be applied as appropriate in this step include:

- Controlled, influenced, uncontrolled (C-I-U) model portraying enterprise state-space parameters and control mechanisms to be incorporated in a multidisciplinary model suite, and integrating (unifying) results of multidisciplinary modeling.
- Program models encompassing POET elements used to simulate performance and estimate outcomes.
- Enterprise architecture (described in Chapter 6) is a valuable representation of the POET elements of the enterprise and data for performance modeling. Its value to management of operations and transformation is enhanced when coupled with the aforementioned simulation methods. It then provides the basic elements of a state-space description of the enterprise that can embody the results of dynamic analyses, and portrays incremental steps in the transformation. As most of the civil agencies are engaged in providing services to the citizen and business, the service content can be represented in the EA as shown in Figure 5.7.

5.8.4 Governance (Strategy, Tactics, R2A2)

Building on the insights into potential operational improvements, risks, and opportunities gained through performance modeling and analytics, implementation issues must be addressed in greater depth. The initial effort toward implementation in this step is governance. Failure to address this step has accounted for impeding numerous promising transformation plans. As most enterprises must have mission continuity during transformation programs, coordination between operations managers and technical program managers, and across both teams at multiple levels, is of the highest priority.

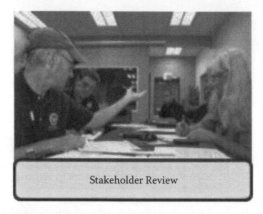

Stakeholder Review

Figure 5.8   The governance model must address requirements and perspectives of multiple stakeholders.

Reference materials, tools, and methods for this critical management step include the process for developing an effective governance plan described in Section 5.6: inclusion of organizational and stakeholder descriptors in the state-space formulation, Chapter 4, and the assignment of specific R2A2 for principal stakeholders and overseers. Specific activities to be performed for this step include:

- Apply the five-step process outlined in Section 5.6 to build effective governance for the transformation.
- Identify the need and opportunities for organizational and institutional changes and their architecting.
- Identify R2A2s of stakeholders and overseers.
- Outline plans, strategies, and authorities for acquisition, integration, and transition to operations.
- Formulate incentives to multiple stakeholders, Figure 5.8.
- Apply the management landscape to align mission performance with enterprise governance and decision style with attention to the nature of the mission and decision processes represented in Figure 5.9.
- Employ the architecting process as outlined in Chapter 6 to incorporate assigned R2A2s for the transformation and continuity of operations.
- Apply agent-based modeling to describe and analyze stakeholder interests, incentives, and reactions to emergent situations.

5.8.5 Implementation (Operations or Transformation Focused)

Implementation is, of course, the critical step in any transformation and will be enhanced through successful completion of the preceding steps to achieve a valid

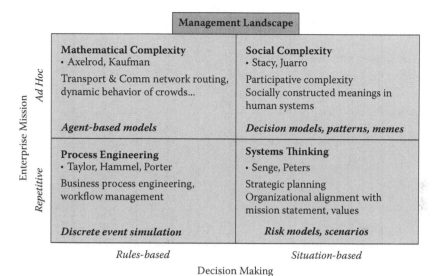

Figure 5.9 (also Figure 4.2)   The enterprise management landscape. (Adapted from D. Snowden and P. Stanbridge *E:CO* 6(1–2): 140–148, 2005.)

architecture, system model, and governance plan. The challenge at this stage is primarily one of management to be informed by those foundational capabilities.

Reference materials, tools, and methods for this implementation step include the governance model and performance models updated to reflect near-real-time implementation activities including acquisitions, staffing, and training. Specific activities to be performed for this step include:

- Implement strategies and tactics for continuity of operations and transformation.
- Implement changes in management structure and mechanisms (controls, influences of agencies and markets).
- Prepare sense-and-respond mechanisms to address uncontrolled aspects and emergent behaviors.
- Engage stakeholders in ongoing collaboration environment informed by real-time data and analytics.
- Maintain program and system models used to relate plans, decisions, and outcomes in a planning and stakeholder "gaming" mode matching program earned value and outcomes against planned progress as shown in Figure 5.10.

5.8.6 Ensure Results/Outcomes (Sense-and-Respond Capability)

Ongoing monitoring of progress in preparation for and during the transformation is needed to prioritize activities, relocate resources, address change requests, and

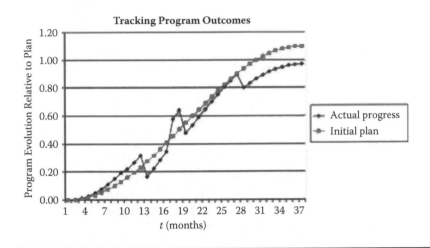

Figure 5.10 Performance modeling applied to a transformation program can simulate progress toward desired outcomes based on activities and key decisions.

report to higher authorities. Attention to internal (mostly controllable) activities and output, and the external environment (mostly influenced and uncontrolled) is required to strengthen program controls, manage influencing opportunities, and respond effectively to uncontrolled events. The timescale of major transformation programs ensures that requirements fixed in the initial stage will change over time, frequently to a significant extent. This reality dictates a dynamic management and resource allocation capability tightly coupled with principal stakeholders.

Reference tools and methods for this more detailed descriptive step include program management tools (PERT, ERP), portfolio modeling and management, environmental scans, third-party evaluation, prototyping test results, and operational evaluation with full stakeholder participation. Specific activities to be performed for this step include:

- Implement sense-and-respond mechanisms to address uncontrolled aspects and emergent behaviors.
- Sense-and-respond to events and disruptions and modify technical and government plans, and perhaps commitments, as required.
- Manage priorities and resource allocation across the program portfolio to accommodate budget restrictions, changing requirements, and emergent behaviors.
- Perform ongoing postimplementation evaluation and modification.
- Stakeholder evaluations, 360 degrees.
- Independent assessments and benchmarking to validate performance against requirements.

5.9 Conclusions

The management of enterprise transformation requires an effective governance environment together with an enterprise transformation process such as ESE/A. The challenges of transformation in complex enterprises, whether single or multiagency missions, are extremely demanding from both the organizational and technical perspectives. As such, they require the integrated efforts of executive managers, operations managers, and technical program managers. The discipline of enterprise dynamics provides a theory and a practical discipline to integrate management processes such as ABM and ERP with EAs and performance models.

A major fraction of the failures in large-scale enterprise transformation programs can be attributed to a lack of attention to the program complexities and challenges across the POET elements that are involved. When applied with effective governance the ESE/A process outlined here provides:

- A disciplined sequential approach to planning, architecting, and carrying out a major enterprise transformation
- A framework to portray the program complexities; the temporal, organizational, and management landscapes; and the various perspectives of stakeholders on the transformation
- A guide to selecting the appropriate qualitative and quantitative models to describe the transformation in sufficient detail to project likely outcomes and emergent behaviors
- A governance structure to incorporate stakeholder perspectives and interest in the outcome, and to reveal reinforcing and conflicting behaviors
- A systems engineering and architecting approach to integrate and harmonize policy, organizational, economic, and technology elements of the transformation

The ESE/A process captures experience and best practices gained over a range of programs. As it is applied going forward, it will evolve to greater levels of reliability and effectiveness. This process will fully support the performance-based and outcomes-oriented transformation guidelines.

Given the role and reach of public and private sector enterprises in national and global operations with extensive partnerships and supply chain relationships, the multiagency model is becoming the dominant mode. The concept of an idealized multiagency governance environment that builds on the capabilities of individual agencies was introduced for very close coupling of technical management with operational management. The architecture-based approach, with a strong focus on mission activities, can deal effectively with the complexity and challenges in achieving this objective. It provides the comprehensive systems perspective of transformation technologies and mission activities needed for successful technical management in a multiagency environment.

Many alternative management options are available during both transformation and operations. Regardless of the option selected, achieving successful multiagency governance dictates a robust governance structure that operates on a comprehensive information base. An architecture-based approach to governance can help to resolve this complexity and the many organizational accountability and control issues that arise in the multiagency environment. This approach will ensure that routine management tasks are integrated across the multiple agencies and responsive to the strategic direction and assigned responsibilities.

The definition of the required level of organizational and technical interoperability is a major factor in multiagency missions. The level of interoperability depends on the specific mission and can vary among the specific activities performed. In any case, it is a strong factor in the assignment of responsibilities and accountability. Specifically, the level of information sharing and provisions for accountability, information quality, privacy, and security are closely tied to the required level of interoperability.

This chapter focused on the contributions of architectures to technical and operations managers. They can play an equally important role throughout all operational phases of multiagency missions. Well-formulated operational activity and service definitions incorporated in the architecture can be used to manage enterprise services and applications, and for financial reporting and performance management. Despite this potential, architectures are underutilized as a strategic planning and operations management resource. The relationships described here to other enterprise planning and management systems suggest how multiple agencies can realize the full potential value of the major investments made in EAs.

References

Bunting, W.J. 2012. Reasoning on uncertain enterprise technology alignment for insight into attainment of enterprise transformation, *Journal of Enterprise Transformation*, 2(1, January): 50–79.

CIO (Chief Information Officer) Council. 2001. *A Practical Guide to Federal Enterprise Architecture* (February).

Clark, T. and Jones, R. 1999. Organisational interoperability maturity model for C2. In *Proceedings of the International Symposium on Modeling and Analysis of Command and Control*. Research and Technology Symposium, United States Naval War College, Newport, RI, June 29–July 1.

DeRosa, J.K. 2005. Thoughts on complex systems and enterprise systems engineering (ESE). Presentation to the Faculty of the College of Engineering and Mathematical Sciences at the University of Vermont, Burlington, December 8.

DoD (U.S. Department of Defense). 1997. *C4ISR Architecture Framework*. Vers. 2.

Executive Office of the President, Office of Management and Budget. 2002. *The President's Management Agenda*.

Hamaker, S. 2003. Spotlight on governance, *Information Systems Control Journal*, 1. http://www.isaca.org/Journal/Past-Issues/2003/Volume-1/Documents/jpdf031-SpotlightonGovernance.pdf

Hoffman, K.C. and Melancon, J. 1988. An information systems architecture for manufacturing/distribution enterprises. In *Proceedings of the ASME Manufacturing International 88*, Atlanta.

Hoffman, K.C., Pawlowski, T., Payne, D., and Zheng, K. 2005. Enterprise business, computing, and information services in a multi-agency environment: A case study in enterprise architect-engineering. In *International Enterprise Distributed Object Computing Conference (EDOC) Workshop, IEEE Computer Society*.

Kuras, M.L. and White, B.E. 2005. Complex system engineering position paper, a regimen for CSE. In *International Council on Systems Engineering (INCOSE) International Symposium*, July.

Malone, T. 2004. *The Future of Work*. Cambridge, MA: MIT Press.

MITRE, C4ISR Architecture Working Group. 1998. *Levels of Information Systems Interoperability (LISI)*. http://www.c3i.osd.mil/org/cio/i3/AWG_Digital_Library/index.htm.

NASCIO, National Association of State CIOs. 2002. *NASCIO Enterprise Architecture Development Toolkit*. Vers. 2.

OECD (Organization for Economic Cooperation and Development). 2004. http://www.oecd.org/daf/corporateaffairs/corporategovernanceprinciples/31557724.pdf

OMB. 1996. (Office of Management and Budget) Memoranda M-96-20, Implementation of the Information Technology Management Reform Act of 1996, April 4, 1996.

OMB FEA (Office of Management and Budget Federal Enterprise Architecture). 2003. *Reference Models Series for Enterprise Architectures (Draft), Performance, Business, Systems and Components, Data and Information, Technology.*

OMB FEA. 2007. *Consolidated Reference Model Document*. Vers. 2.3.

Player, S. and Keys, D.E. 1999. *Activity-Based Management*. 2nd ed. Hoboken, NJ: John Wiley & Sons.

Snowden, D. and Stanbridge, P. 2005. *E:CO* 6(1–2): 140–148.

Sussman, J.M. 2002. Representing the transportation/environmental system in Mexico City as a CLIOS. Presentation at the Fifth US-Mexico Workshop on Air Quality. Ixtapan de la Sal, Mexico.

U.S. Congress. 2000. Clinger–Cohen Act (PL 104-106).

Zachman, J.A. 1987. A framework for information systems architecture. *IBM Systems Journal*, 26(3): 276–292.

ENTERPRISE MODELING APPROACHES AND APPLICATIONS

II

Section II presents a series of Enterprise Systems Engineering and Architecting (ESE/A) case studies based on published papers and invited presentations for MITRE technical exchanges, The case studies describe a range of approaches to analyses of the complex dynamics in transforming specific enterprises as they acquire new operating capabilities.

The enterprise challenges addressed involve a mix of public and private stakeholders, some in even more complex multiagency operations, with an emphasis on major transformational initiatives where government policies and programs play an important role. These range in scale from such focused operations as command and control centers that deal with security, to major sectors of the U.S. and global economy such as energy systems, healthcare, and the monitoring and management of international trade.

The models applied to these enterprise challenges are drawn from multiple disciplines that approach Enterprise Dynamics from different perspectives—technical, economic, and management. Collectively, they support the steps in the Holistic ESE/A process described in Section I, Section 5.8.

The case studies of complex systems and megasystems applications cover a wide range of public and private enterprise domains at multiple scales, all with a focus on mission performance. They have both employed and contributed to the ESE/A process and concepts described in Section I, and illustrate the tailoring of the process to address the unique complexities of the application domains.

Each chapter describing a specific ESE case study touches on the following topics drawn from the ESE/A process as applicable:

1. **Description of the enterprise**

   The description of a significant undertaking by an agency or larger set of public–private entities defines the enterprise that is the subject of a case study. The description includes the specific undertaking of the enterprise and its purpose, along with the responsible operational entities and stakeholders that have an interest in its performance.

   The case studies emphasize public–private undertakings of national importance involving security, the economy, and the environment. Accordingly, the enterprise will involve a mix of stakeholders whose activities and interests are interrelated in complex ways.

   The enterprises, or "undertakings" range from very specific operations for a narrowly defined, or focused, activity to regional and national initiatives that involve major sectors of the U.S. economy. For the large scope and scale enterprises, the case studies emphasize descriptive and analytical methods that illuminate complex interrelationships that must be dealt with by a large number of stakeholders in decisions and actions taken from their individual perspectives. The narrower, more focused, enterprises have well-defined objectives that can include definition of those situation-dependent decision processes and methods.

2. **Statement of the enterprise transformation challenges**

   Government and commercial operations increasingly require the integration of a complex mix of business activities—and services from multiple public and private organizations—to perform missions and deliver services of high national importance. When these operations are to be modernized to reach higher levels of performance through new business processes supported by advanced technology and information systems, they present unique engineering and governance challenges that demand increasingly powerful analytical capabilities.

   The statement of capabilities to be achieved in the transformation of the enterprise is the starting point for ESE/A. The improved capabilities may be targeted on the interrelated elements of the overall enterprise, or may be directed at a specific entity or activity within the enterprise that requires the full enterprise perspective.

   Complexities of each application are outlined along with the approach taken to deal with technical and system complexities, emergent behaviors, and major uncertainties.

3. **Characterization and representation of the scope and structure of the enterprise in its operating environment**

The initial state of the enterprise is described, along with the new and/or improved capabilities and performance objectives to be attained in the transformation. The nature of the enterprise and the products and services it delivers are described along with a characterization of the operations ranging from repetitive activities to ad hoc reactive and proactive activities that are highly situation dependent.

A state-space description of the enterprise indicates the parameters of its phased transformation from the current state to the future state that must be modeled. Control parameters that may be applied directly or influenced by stakeholders are also defined, along with important factors in the external environment that can affect the enterprise.

Major features of the Enterprise Architecture (EA) are described to represent the mission/business activities performed, the resources and services utilized, and supporting information systems including networks, software, and computing.

The methods used to characterize the enterprise and represent its scope and structure will describe the specific policy, organizational, economic, and technology (POET) elements involved in the transformation.

4. **Strategies and technologies examined for the transformation**

The strategies and options to be considered will generally be constructed around a technological strategy that promises significant performance improvements. Transformation activities will center on the acquisition of technologies and systems, along with the associated organizational change, process change, and policy changes.

5. **Analytical approach—models applied to analyze performance and transformation dynamics**

The logic for selecting and applying specific models is outlined. The analysis will map the transformation using a state-space representation from the current situation to the transformation objective. Models and methods will be based on the specific emphasis placed on each of the POET elements in the state-space description.

The models are applied to map the state-space trajectory for transformation of the enterprise with specific attention to control parameters, influences, and uncontrolled perturbing events both internal and external to the enterprise.

6. **Governance**

In the final analysis, successful transformations are dependent on effective management and stakeholders working together toward enterprise objectives. Stakeholders need to be involved in specific aspects of the planning, transformation and operational phases. The Roles, Responsibilities, Authority, and

Accountability (R2A2) of key stakeholders may be defined based on the characterization of the enterprise and its performance challenges.

7. **Results and recommendations to achieve operational excellence**

Application of the ESE/A process leads to a set of results and recommendations. These will address the overall enterprise objectives and the integration of activities, as well as the responsibilities of specific stakeholders, public and private.

8. **Conclusions**

The strengths and weaknesses of the ESE/A analyses are discussed at the end of each Section II chapter. The emphasis is on areas requiring greater attention and those that require additional research and development on tools and methods.

The analytical models described in Section I and utilized in the Section II case studies were mapped with respect to their coverage of POET elements in Figure 4.7 presented earlier. The coverage of specific models can vary depending on their scope and purpose. The mapping is representative of the original emphasis of these methods; however many successful examples of more comprehensive coverage can be found in the literature.

The perspective on the strengths of multidisciplinary models illustrated in Figure 4.7 indicates that no single model can effectively address the full POET elements of enterprise transformation. For comprehensive coverage of the "enterprise state-space" integrated, unified, hybrid methods are essential.

The contents of the chapters in Section II represent case studies and analyses to address very specific issues and national challenges as described below. They draw upon tailored versions of the ESE/A process and tools and methods applicable to the issues and complexities. Some of these employed the ESE/A process, others informed the evolution of the process.

The case study applications are organized in relation to the defined scale of the enterprise starting with single agencies and corporations, progressing through more complex mission operations to multiple agencies and challenges of national scope with significant impacts on the economy. This range of scope and complexity of enterprises addressed (as significant undertakings) is summarized as shown in the following Table II.1.

Following is a summary of the case study chapters in Section II that describe ESE/A approaches across various scales of the enterprise:

Chapter 6: Simulation of Enterprise Architecture for a Business Strategy

The coupling of enterprise architecting with Enterprise Dynamics in this chapter is applied to strategic planning at the corporate level. The results of alternative investments of discretionary funds in business development and R&D are analyzed.

Table II.1   Scope of Enterprise Addressed in Section II Case Studies

Case Study Chapter	Single Agency or Corporation (Applicable at multiple scales)	Complex System or Operation	Multiagency Undertaking	National Challenge (involving public and private sectors)
6. Simulation of EA for a Business Strategy	X			
7. Reasoning for Agency Transformation	X			
8. Optimal Control Modeling of an SE Process for Transformation		X		
9. Hybrid Modeling of an Air & Space Operations Center		X		
10. Nuclear Waste Management Program		X		
11. International Trade and Commerce			X	
12. Energy & Materials Systems Planning and Analysis				X
13. Modeling the Nation's Healthcare System				X

This case study represents a major step forward in the foundational discipline that supports ESE/A through the concept of dynamic architecting. The application described implements a dynamic model and simulation with the comprehensive structure of an EA as defined by federal and commercial EA guidelines. The approach overcomes a major deficiency in the static nature of most architectures by joining this widely-practiced enterprise planning and management method with Enterprise Dynamics.

Chapter 7: Reasoning on Technology Uncertainties for Enterprise Transformation
This chapter describes a method for reasoning about the likelihood of attaining specified desired performance dynamics resulting from emerging dynamics, and uncertainties within an enterprise or its environment. A basic principle is that mission performance occurs from the individuals understanding the dynamic interactions of people, process, and technologies. Individuals reason on dynamic interactions producing insights on likely performance attainment given the interactions. Individuals increase their respective understanding and their subsequent decisions and actions changed because of their improved collective understanding.

The method uses evidential reasoning and Multientity Bayesian Networks (MEBN) to compose arguments about the relationship of technology initiatives to strategic outcome attainment. A federal program example illustrates the method.

Chapter 8: Optimal Control and Differential Game Modeling of a Systems Engineering Process for Transformation
This case study describes a unifying analytical framework for modeling Enterprise Dynamics in the ESE/A process, across a range of enterprise and program types and is demonstrated on a specific transformation program. The framework is a control theoretic formulation that uses state and control parameters relating to the enterprise, stakeholders, and the environment with which the enterprise interacts. The framework expresses the individual self-interests of stakeholders and environmental players as a differential (dynamic) game, which is explored and interpreted in terms of game-theoretic Nash equilibrium solutions. It provides a mathematical basis for relating and/or integrating the diverse ESE/A modeling methods.

Chapter 9: Hybrid Systems Dynamic, Petri Net, and Agent-Based Modeling of the Air and Space Operations Center
This chapter describes an innovative enterprise systems engineering effort to model the policy, organizational, and technical aspects of a mission critical national defense operation. The application uses hybrid systems dynamics, Petri net, and agent-based multiscale modeling to understand the effect of operator-environment interaction and the global environment on Air and Space Operations Center (AOC) processes. The AOC process model is linked (e.g., critical event process time and probability of errors) to a global-environment model that is driven by the political landscape in which the AOC operates.

Chapter 10: Nuclear Waste Management Strategic Framework for a Large-Scale Government Program
This case study presents an enterprise-wide framework using the hybrid approach of a process-based materials flow model coupled with a systems dynamics influence diagram, or causal loop diagram. The objective of the work is to provide system insight into the U.S. Department of Energy's (DoE) responsibility for environmental cleanup of legacy nuclear waste. The focus is on the Savannah River Site and all activities carried out in this enterprise from the receipt of nuclear materials through their processing to the shipment of materials in forms suitable for safe long-term storage. The framework is used for exploring policy options, analyzing plans, addressing management challenges and developing mitigation strategies for DoE Office of Environmental Management (EM). The sociotechnical complexity of EM's mission compels the use of a qualitative approach to analysis to complement a more a quantitative discrete event modeling effort. We use this analysis to drive scenarios for the model, pinpoint pressure and leverage points, and develop a shared conceptual understanding of the problem space among stakeholders. This approach affords the opportunity to discuss dynamic phenomena in enterprise operations over a 25-year time horizon using a unified conceptual perspective and is also general enough that it applies to a broad range of capital investment/production operations problems.

Chapter 11: International Trade and Commerce: Enterprise Systems Engineering and Architecture in a Multiagency Environment
The objective of this case study is to formulate and demonstrate a comprehensive planning framework for ESE/A in a complex international enterprise that is essential to the global economy. The framework, an integrated Enterprise Systems Engineering (ESE) Workbench, is designed for multinational and multiagency enterprises, public and private, engaged in international commerce.

Chapter 12: Energy and Materials Systems as an Enterprise Systems Engineering Application: Planning and Analysis for the Economy's Infrastructure
This case study application provides an example of a large-scale ESE challenge to the private sector and government, crossing major materials and energy-related sectors of a nation's economy. Major transformational initiatives have been proposed to deal with resource depletion and environmental challenges. The energy and materials systems underlying the physical infrastructure of national economies are complex and addressed here through combined technical and economic analytical methods.

Chapter 13: Modeling the Nation's Healthcare System as a Dynamic Enterprise
This case study provides an example of a government ESE challenge to transform healthcare to provide greater access at sustainable cost with effective outcomes. Health sectors account for 18% of the U.S. economy with roughly half of that

funded through government programs. This application spans POET aspects of the challenge at multiple scales ranging from demographics through specific diagnostic and therapeutic services, the structure of healthcare sectors and interactions with other sectors of the economy, to the overall economy. A framework is presented that helps to integrate analytics in this data-rich, but information-poor environment.

Epilogue: Enterprise Systems Engineering and Architecting—Lessons Learned and the Road Ahead

The Epilogue reinforces descriptive Enterprise Dynamics as an essential part of enterprise planning and analysis for the acquisition and implementation of transformational technologies and processes to improve mission performance (ESE). This dynamic perspective of POET aspects of the enterprise is a central feature of the ESE/A process. This sourcebook is a snapshot of the current state of an emergent field of national import; the authors are motivated by the opportunity to contribute an assessment of this current state and a platform for further research and operational applications.

Chapter 6

Simulation of Enterprise Architecture for a Business Strategy

Christopher G. Glazner[*]

Contents

[*] Adapted from Glazner, C.G. Enterprise Transformation Using a Simulation of Enterprise Architecture. *Journal of Enterprise Transformation*, 1(3): 231–260, July 2011. With permission from Taylor & Francis Ltd, http://www.tandf.co.uk/journals

6.1 Introduction

Today, despite the efforts of researchers in many fields over the past half-century, the design and management of enterprises remains as much art as science. The complex structure and behavioral dynamics of enterprises makes it difficult to untangle the relationship between form and behavior; changes to one aspect of an enterprise's structure, incentives, or strategy can affect its behavior in seemingly unrelated areas at distant points in time. An enterprise is composed of many different elements, such as the business plan, organizational structures, processes, and technologies, and all of these components interact, often in dynamic ways that are difficult for an individual or group with limited visibility and cognitive capacity to anticipate. Nonlinearities, inertia, delays, and feedback in the system all contribute to the difficulty in understanding how complex enterprise behaviors are influenced by the "design" of the enterprise that produced them. It is exceedingly difficult for enterprise leaders to anticipate how any changes to their enterprise's form may affect its behavior without tools to help them analyze the behavior and its drivers.

Imagine the task of a chief executive officer, faced with a shifting business environment challenging the existing business model. There are a host of questions that may be asked in such a situation:

- How can the enterprise be retooled to capitalize on a newly developed business model?
- Will a new business model require changes to the enterprise's organization, processes, or knowledge requirements?
- How can a new organizational form and incentives be developed that will be responsive to new customer demands or more quickly take advantage of new technologies in its products and processes?
- How could alternative proposed forms of the enterprise be compared to each other? How could trade-offs be assessed?

Currently, enterprise leaders have few tools available that can help them wrestle with questions such as these that concern the enterprise's architecture, its fundamental design. Most such major decisions concerning an enterprise's architecture are made without the aid of a tool that allows experimentation to test hypotheses. Previous practical efforts to develop tools and processes to conceive and analyze aspects of the enterprise, such as business process re-engineering (Hammer and Champy 1993) and value stream mapping (Womack and Jones 1996) have relied heavily on "brown-paper walling" or "white boarding," employing sticky notes and hand-drawn lines to convey new organizational structures and processes. More recent efforts have simply used digital versions of this static "boxes and lines" approach to modeling the enterprise's structure. Such approaches do not create dynamic models that can be subjected to hands-on experimentation, however. They do not embrace the use of tools and concepts from the field of e-architecting. What is required is the capability to bridge the gap between descriptive approaches to enterprise design and quantifiable models from which results can be collected and analyzed (Fowler 2003). Without the ability to analyze new architectures, there cannot be a formal reliable process for designing (or redesigning, or evolving) the enterprise (Levitt 2004).

6.2 Enterprise as a Complex System

Unfortunately, an ideal comprehensive model that captures every facet of an enterprise's design and behavior would be extremely difficult, if not impossible to create in practice. As demonstrated in Chapter 3, enterprises are very complex systems, many orders of magnitude more complex than the largest models that modern computers can handle (Simon 1990). Enterprises are both structurally complex, in that they have a great number of interconnections, as well as behaviorally complex, in that the behavior of the system cannot be understood or anticipated by study of the constituent parts; its behavior is more than the sum of its parts. There are many factors that make enterprises complex. Enterprises are

- Sociotechnical systems, combining "hard" technical elements as well as "soft," cultural and organizational elements
- Highly interconnected with many feedback loops, inertia, and delays
- Filled with autonomous people all making local and perhaps not strictly rational decisions
- Capable of adaptation
- Embedded in a constantly shifting, open environment

The complexity of enterprises makes creating the ideal crystal ball simulation model almost impossible. A predictive simulation of a complex system that seeks to model all inputs, outputs, and interactions is a futile endeavor. The data for such models are often approximated, and the full nature of all interactions is not known. Enterprises exhibit chaotic behavior: they are stochastic and sensitive to small perturbations to their conditions that make accurate prediction close to impossible (Dooley and Van de Ven 1999). Sterman (1991, pp. 209–229) argues "It is simply not possible to build a single, integrated model of (a complex system), into which mathematical inputs can be inserted and out of which will flow a coherent and useful understanding of world trends." A fully detailed, predictive model of a complex system would necessarily be as complex as the original system.

The key to modeling complex systems such as an enterprise is to properly abstract them. Modelers must abstract from the complexity of the enterprise in a way that helps to highlight critical interactions and relationships that drive behaviors of interest, while ignoring other interactions that do not contribute to a systems-level understanding of the enterprise. Herbert Simon (1990) argues that "intelligent approximation, not brute force computation, is still the key to effective modeling." For many problems, the answers that are needed do not require a highly detailed, predictive model, but rather one that is capable of understanding general trends, paths, and steady states. Rather than focus on models that try to predict the chaotic behavior of enterprises, modelers and leaders should instead seek out "organization-specific generative models that can explain how the chaotic behavior came about in the first place" (Dooley and Van de Ven 1999). Enterprises must be understood in terms of their dynamic behavior, and models must be designed to capture patterned sequences of events driven by the organizational design (Abbott 1990). Enterprise modelers should not concern themselves with how to build more detailed models, but instead with how to develop abstractions of enterprises in a way that allows more insight into their behavior, and ultimately do a better job of conceiving and managing them.

6.2.1 Abstracting Enterprise Complexity

One approach to developing an abstraction of the enterprise, as explained in Chapter 3, is the practice of *enterprise architecting*. Enterprise architecting holds that the enterprise can be both understood and designed by employing the construct of enterprise architecture as a unifying conceptual framework. The enterprise architecture is a documented abstraction of the fundamental organization of an enterprise as a dynamic holistic system with nonlinearly interacting components. The architecture of an enterprise is an abstraction of its essential features, rather than a complete detailed description of its design. In order to help enterprise architects develop these abstractions, enterprise architecture frameworks (such as the policies, organizations, economics, and technology (POET) framework used in this sourcebook) can serve as a starting point in establishing scope and identifying key abstractions, boundaries, and interactions within an architecture. Each framework identifies multiple *views* that can be used to decompose the architecture from different perspectives, such as strategy, organizational structure, processes, and information technology. These views are interconnected together as a system.

Unlike the theoretical literature on organizational design and architecture that remains highly fragmented and multifaceted, an important immediate benefit of the enterprise architecture reference frameworks is that they provide a practical way for defining and decomposing enterprises as an interconnected system, relating the disparate components of the architecture together. Enterprise architecture frameworks are intended to serve as a unifying platform for many different disciplines of enterprise design and study to allow their combined and coordinated application, rather than their individual disjoint application. They do not provide any theoretically grounded guidance of their own to guide the architecting process, but instead unify the theories developed in disparate research areas such as organizational science, management science, and business process design in order to achieve this task. They provide well-defined workable representations of aspects of the enterprise that can then be analyzed by employing a number of methods to help understand and create enterprise architectures.

Enterprise architecture frameworks provide a potentially useful way of simplifying and abstracting an enterprise's complexity by decomposing it into its major constituent components, each representing a discrete view, and by linking them together into an interconnected whole, by utilizing associated theory. The resulting enterprise architectures, however, are static representations and present limited opportunity for quantitative analysis of the underlying dynamic enterprise architecture. It is thus difficult to use an enterprise architecture framework alone to answer a question concerning the enterprise's potential behavior. To answer such questions, a simulation model of enterprise behavior is needed. The enterprise architecture, however, may be used to provide the necessary abstractions, scoping, structure, and interactions necessary to create a simulation model to answer the question.

6.2.2 *Hybrid Simulation Modeling*

The complexity of enterprises requires that they be abstracted into different subsystems in order to be understood. Each subsystem presents a new perspective on the enterprise, complete with its own context. For the modeler attempting to model behaviors this presents a challenge: many problematic behaviors span multiple perspectives, and there is no one single simulation modeling methodology that is capable of simulating each perspective in its own context (Mingers and Gill 1997). The tools used to simulate the behavior of an incentivized collection of people within an organization are very different from those used to simulate the execution of a manufacturing process or top-down strategic resource allocation decisions. Each simulation methodology has its own strengths and weaknesses, based on the assumptions and mechanisms that it uses to simulate the system at hand. For this reason, it has been argued that a cross-disciplinary approach to simulation modeling of enterprises be taken, employing a portfolio of models from different fields and for different purposes, allowing the individual models to be compared, contrasted, and critiqued (Sterman 1991).

Others have gone on to argue that using a portfolio of stand-alone simulation models does not accurately convey the system's dynamics, and that a hybrid multimethodology approach to simulation should be used (Mingers and Gill 1997; Rabelo et al. 2005). In a hybrid simulation model, multiple submodels employing different simulation methodologies are interfaced with each other such that the execution of one submodel can be used as the input of another, forming a system of interacting subsystems. In recent years, this hybrid approach to modeling complex systems has gained traction in some areas of enterprise modeling, such as supply chains (Scheritz and Größler 2003; Rabelo et al. 2007), production planning (Venkateswaran and Son 2005), and manufacturing decision making (Rabelo and Speller, Jr. 2005). These hybrid simulation models have employed system dynamics, agent-based models, and discrete event simulations to capture different perspectives of a system, and have applied each methodology to areas where it is the best method to describe and understand the system's behavior.

To date, the few hybrid models of enterprise operations have been fairly modest in scope. There has not been a formalized approach to determining the boundaries between submodels in the hybrid models or the pathways of their interactions. Previous hybrid modelers have taken an ad hoc approach to boundary setting between simulation submodels. Ad hoc approaches will become increasingly problematic as the scope and complexity of hybrid models increases. Furthermore, each time a new model is required to capture a new aspect of the same enterprise's behavior, a new effort to abstract the portions of the enterprise being modeled is required. This can prove to be a very substantial amount of work, and may also provide obstacles to communicating the boundaries and scope of the model.

6.2.3 An Enterprise Architecture-Based Hybrid Simulation Methodology

Simulation models of enterprise behavior can take advantage of an initial definition of the existing enterprise architecture in order for them to capture an adequate

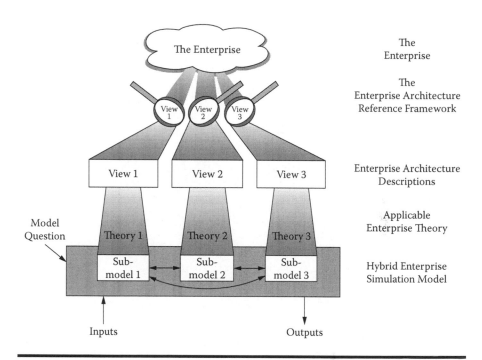

Figure 6.1 The method for creating hybrid, enterprise architecture-based simulation models. (Reprinted from C.G. Glazner, *Journal of Transformation*, 1: 231–260, 2011. With permission.)

abstraction of the enterprise as it exists and thus guide their boundaries. Using such an approach, the views of the enterprise architecture can be used to define the boundaries of the submodels. Interactions among submodels can then be modeled using the interview interactions in the enterprise architecture, by employing known theoretical propositions. The use of the existing enterprise architecture as a reference framework ensures that the hybrid model is consistent with the developed architecture of the enterprise and permits the model to be more easily understood and communicated to enterprise stakeholders.

This chapter proposes an approach outlined in Figure 6.1 for the creation of hybrid, enterprise architecture-based simulation models that can be used to analyze enterprise behaviors driven by the enterprise's architecture. At the top of the figure, the complex, real-world enterprise is represented as a cloud. An enterprise architecture framework, shown as a set of lenses, is used to focus and abstract the enterprise to produce the enterprise architecture, shown as a set of three boxes representing three possible views within the enterprise architecture corresponding to the framework used. This enterprise architecture is a static representation of the enterprise, and primarily serves a communicative role.

When enterprise leaders have a question about how their architecture may affect the behavior of the enterprise, a model can be constructed using the question and the boundaries and structures identified in the enterprise architecture. Theoretical constructs from pertinent bodies of enterprise theory are then used to guide the creation of the submodels using a simulation methodology that is matched to the representative behavioral dynamics of each view.

There are many potential benefits of such an approach to simulating enterprises. By modeling multiple perspectives of the enterprise simultaneously, hybrid simulations can be created that do a much better job of analyzing behaviors driven by interactions across these perspectives. Without a hybrid approach to simulation, these cross-enterprise interactions cannot be effectively simulated, and enterprise leaders will not have any tools at their disposal to investigate the effects of architecting choices or to explore high-performing variations to the architecture. Without such an analysis capability, enterprise architecting and management will remain far more art than science.

Using an enterprise architecture to guide the development of the simulation also has the important benefit of providing a pre-existing abstraction of the enterprise for the modeler to use, eliminating the need to structure the model in an ad hoc fashion based upon the modeler's independent investigation. Using this existing abstraction makes model component reuse a possibility, as submodels will all use the same boundaries and interfaces. This allows an enterprise to build up a library of submodels that can be more quickly interfaced to create hybrid models to answer new questions. Simulation models based on enterprise architecture also aid communication of the model to stakeholders familiar with the enterprise architecture.

6.2.4 An Application of Hybrid Simulation Modeling of EA: The TechSys Case Study

The approach outlined above for creating a simulation of an enterprise's dynamics to solve real-world enterprise challenges shows promise, but demands a practical application before it should be considered worthy of further pursuit and development. To this end, the approach was applied to simulate the dynamics of "TechSys,"* a multibillion-dollar aerospace/defense sector company undergoing significant strategic, organizational, and process change. The TechSys case study provided an outstanding opportunity to apply this approach in a complex realistic setting. Using the model of TechSys dynamics produced the following approach: TechSys was able to identify key strategic, organizational, and process mismatches that prevented it from realizing significantly increased revenues. This dynamic analysis approach helped uncover changes that could be made to the enterprise architecture, which although not costly to the enterprise, yielded the potential for millions of dollars worth of improved performance.

TechSys was in a good position to benefit from the creation of a hybrid enterprise architecture simulation model. It had recently undergone a series of changes

---

* "TechSys" is a pseudonym used to protect the identity of the organization.

to its enterprise architecture and was trying to understand the connection better among its enterprise architecture, its strategy, and enterprise performance. In order to create this model, however, much work was required. First, the problems that TechSys faced would need to be clearly articulated in a way that could be addressed using a simulation model. There were a number of potentially valuable areas that could be explored using the hybrid simulation modeling approach, but the key question that TechSys wished to address with the simulation model was how to better identify, manage, and overcome barriers to technical and business collaboration among its operating units.

TechSys had previously operated in the mold of a holding company, with a number of operating units in different, loosely related markets. As part of a strategic initiative, TechSys was rearchitecting itself to benefit from potential synergies among its operating units. New strategies, organizational designs and incentives, and processes had been developed to execute this strategy. The key question, however, was whether the newly rearchitected TechSys would be able to meet its strategic objectives as it had been redesigned. The simulation model was developed to answer this question, taking into account the dynamic interactions among multiple enterprise architecture views.

6.2.4.1 Identifying and Bounding the Root Problem

After the general topic for the simulation model had been chosen, the first step to problem articulation was to more concretely define and bound the topic into an addressable problem. The chosen topic, "collaboration among operating units," would need to be refined to determine what aspects of collaboration should be considered and reviewed to determine that the area of interest was truly collaboration, and not something broader.

There are many ways collaboration can occur, such as sharing market information, sharing best practices, or exchanging personnel. In the case of TechSys, discussions uncovered that they were most concerned about how operating units could work collaboratively to pursue new business opportunities, and how this collaboration could be used to increase the competitive position of the greater enterprise. New business pursuit encompasses the activities ranging from the development of ideas for new business opportunities through everything required to submit and win a new business contract. During the most recent round of rearchitecting, TechSys attempted to create an enterprise architecture that could facilitate collaboration between the operating units when pursuing new business opportunities, allowing them to win new business they could not have won without collaboration.

TechSys uses the term "synergy growth" to describe the kinds of "business opportunities pursued jointly between two or more operating units that increase the competitive position of the enterprise." Synergy growth is growth that comes when two or more operating units work together to increase effectiveness in existing markets and pursue new markets. Synergy growth is contrasted with "organic" growth, which is growth that advances the existing business of a single operating unit without collaborating with any other operating unit. The combination of both synergy growth and organic growth leads to total new business growth for the enterprise. The challenge that the enterprise faces is twofold: (1) how does it enable synergy growth between operating units, and (2) what is the right balance between synergy and organic growth to maximize total new business growth across the enterprise, assuming synergy growth is possible? The enterprise has limited resources, and they must be allocated between pursuing synergy and organic growth.

Given this, the original topic of "collaboration between operating units" has evolved substantially toward a deeper understanding of root problems of importance to TechSys. The goal of the enterprise is not simply to maximize synergy growth (increase collaboration); it is to maximize the total new business growth across the enterprise. This can be achieved through increasing synergy growth, organic growth, or a combination of the two. TechSys must make decisions regarding the allocation of resources to growth, but it currently has no tools or process that would allow it to determine the effectiveness of investments in promoting synergy growth and no way to understand the trade-off between synergy growth and organic growth. One of the intended goals of the TechSys simulation model is to provide this capability.

To be genuinely useful to TechSys, the simulation model focused on the total pursuit of new business opportunities, rather than a single component of growth. The enterprise architecture-based simulation model must help identify levers in the architecture to enable synergy growth as well as show the outcome of strategies for resource allocation toward synergy and organic growth to achieve maximum total growth.

The dynamics of TechSys's architecture for new business opportunity pursuit and capture are complex, but the problem area can be well bounded for purposes of modeling. The dynamics of new business opportunity pursuit and capture are strongly driven by strategy at both a divisional and operating unit level, with local decisions made in an organizational context, following established processes, dependent on aligned knowledge requirements, supported by information technology, and bounded by external constraints. The simultaneous interaction of all of these factors across the enterprise architecture can make for complex behavior that cannot be analyzed without the aid of a model of the architecture. Fortunately, the problem area is neatly bounded by a handful of processes and a specific organizational structure with clear inputs and outputs. For all of these reasons, this problem was well suited for evaluation using a hybrid, enterprise architecture-based simulation.

The key questions for TechSys surrounding the pursuit and capture of new business opportunities include:

- Can TechSys achieve its growth goals (both synergy and organic growth) given its current enterprise architecture with constrained resources dedicated to growth?
- How sensitive is the architecture to changes in resource allocation?

- What changes can be made to the architecture to improve growth opportunities given constrained resources?
- What combination of inputs should be used to best grow the enterprise?

6.2.4.2 Identifying Inputs and Outputs

After clearly stating the problem, the next step of problem articulation is to identify the key input parameters and the output of the simulation model. It is usually easier to determine the output of the model first, as there is often a clear idea of some quantity that must be maximized or minimized. This was true for TechSys; they sought to track the revenue and profits that arise from the capture of new contracts. Revenue and profits in this model can come from two sources: organic growth from a single operating unit, or synergy growth arising from a contract awarded to two operating units cooperating on a contract.

Identifying the inputs for the model required an examination of the controls that TechSys management uses to influence the enterprise's ability to pursue and capture new business, such as funding, headcount, and other resources. It also required a review of the process to identify any other potential inputs that may not be in current use by TechSys. At an operational level, there were over a dozen possible inputs identified that had some impact at some point in the processes. Working with TechSys stakeholders, the large list of possible inputs was distilled down to five strategic resource inputs that management had direct control over and used to influence the enterprise. After iterative discussion with senior leadership, the following model inputs were used:

- Percentage of new opportunities pursued with a synergy component: This tracked the mix of new business development opportunities pursued. The lower this fraction, the more operating units would continue to pursue business independently of each other.
- Discretionary budget: The amount of money to be allocated to either internal research and development (IRAD) or business development (known as bid and proposal).
- Percentage of discretionary budget: The amount assigned to bid and proposal (with the remainder going to IRAD).
- Indirect marketing budget: The amount that can complement the bid and proposal budget.
- Headcount: The number of people involved in associated processes, tied closely to budgets.

Figure 6.2 is a notional "black box"-level depiction of the simulation, showing the set of input parameters on the left that are fed into a "black box" with a transfer function T which is a function of the input parameters. The output of the

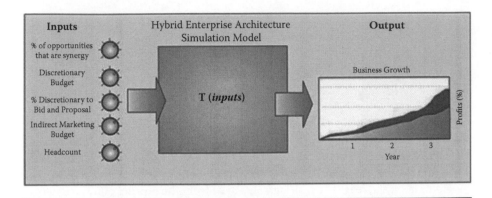

Figure 6.2 Conceptual "black-box" level depiction of the TechSys simulation model. (Reprinted from C.G. Glazner, *Journal of Transformation*, 1: 231–260, 2011. With permission.)

transfer function, synergy, and organic profits due new contracts won, is shown on the right of the transfer function, shown as a stacked cumulative graph.

6.2.4.3 Framework Selection

For this modeling effort, a new enterprise architecture framework developed by Nightingale and Rhodes (2004) was used. The Nightingale–Rhodes Enterprise Architecture Framework (NREAF) employs eight highly interacting views (strategy, organization, process, technology, knowledge, products, services, and external/policy). It is more closely aligned with holistic systems frameworks such as POET than traditional enterprise architectures (EA) frameworks such as the Department of Defense Architecture Framework (DoDAF) or The Open Group Architecture Framework (TOGAF). This framework was used because it was created with the intent to serve as an EA relevant to strategic decision makers affiliated with the CEO, rather than traditional information technology and systems decision makers in the CIO's office.

6.2.4.4 Key Structures and Behaviors

The next step of problem articulation was to link the problem to the architectural structures and dynamics of the enterprise by identifying key structures and behaviors in the enterprise architecture that are relevant to the problem. The problems that TechSys faces balancing its new business pursuit capture are linked very deeply to its enterprise architecture, and are fundamentally dependent on facets of the architecture described by the strategy, process, organization, and knowledge views within the NREAF.

Figure 6.3 gives a high-level, notional view of the important factors and structures from the strategy, organization, process, and knowledge views associated with

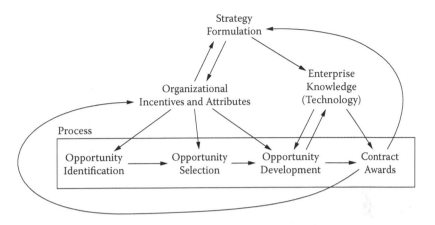

Figure 6.3 **A conceptual diagram of relationships between architectural factors related to the pursuit and capture of new business opportunities. (Reprinted from C.G. Glazner, *Journal of Transformation*, 1: 231–260, 2011. With permission.)**

the pursuit and capture of new business opportunities and how they relate to each other. This very simple sketch highlights how the simple linear process of opportunity development (identification to selection to development to contract award) is part of a larger process across the enterprise with feedback. At the top of Figure 6.3, strategy formulation is shown driving both organizational incentives and characteristics as well as the pursuit of new knowledge (technology). It receives feedback from both the performance of the organization as well as the ultimate performance of the process in the form of awarded contracts. The organization's incentives and attributes drive every stage of opportunity pursuit, from identification to selection and then development. Furthermore, the organizational incentives determine the extent to which the operating unit will participate in developing synergy opportunities and collaborate with others. The organization also receives feedback from the performance of the process in the form of contract awards before it arrives in the form of a new strategy. The success of the new business opportunity in the development stage is dependent on the knowledge and capabilities of the enterprise. After the work is completed and if it has been internalized, the organization has then gained experience that will aid future development efforts. That experience also establishes a reputation that increases the likelihood of future contract awards.

6.2.4.5 Form a Dynamic Architectural Hypothesis

After clearly articulating the problem for the simulation model of TechSys's enterprise architecture, the next step in the hybrid enterprise architecture modeling process is to further scope the boundaries and function of the model by creating a dynamic architectural hypothesis. The dynamic architectural hypothesis, as defined by Sterman (2000), is based on an initial insight into the problem that explains its dynamics in terms of the enterprise architecture. Usually, the enterprise's stakeholders will have some opinions as to the cause of problematic behaviors or barriers to achieving a performance goal. The dynamic architectural hypothesis is a chance to articulate these initial beliefs.

The dynamic architectural hypothesis for the TechSys simulation model was developed with input from the stakeholders, with particular input from the director of enterprise strategy, who was actively leading a team at TechSys working to identify and mitigate barriers to synergistic cooperation between operating units at the time. There was a concern at TechSys that the current growth goals established in its strategy were unattainable using only organic growth; new synergy growth between operating units that would enable expansion into new markets would be necessary. Given recent performance of the enterprise, most felt that the existing architecture was not capable of realizing the necessary synergy growth between the operating units in sufficient quantity to meet its growth goals without the removal of several barriers to synergy, many of which were a part of the enterprise architecture. In light of this environment, the following dynamic architectural hypothesis was proposed for the TechSys simulation model:

> Given its current enterprise architecture, TechSys will be unable to meet its business growth and associated profitability goals by utilizing the pool of budgetary resources that are available for pursuing new business growth opportunities. That is, the existing architecture is hypothesized to have a constraining influence on the company's capacity to capture the new business opportunities it has targeted. Hence, the existing architecture must be modified in order for the company to take advantage of the synergistic business growth opportunities facing its various business units in order to achieve the new companywide business growth goals.

The motivating hypothesis or central strategic question posed above can be evaluated by running a range of feasible inputs (limited budgetary resources) through the simulation model and observing if any combination of inputs can cause the expected value of TechSys's business growth or profits to meet its established goals. If the model can be shown not to result in the new business growth or profitability targets, under any combination of the available types of budgetary resources and how they are allocated, then the maintained hypothesis (i.e., the current enterprise architecture has a constraining influence on the achievement of the targeted business goals) cannot be rejected. If the outputs of the simulation model can meet

TechSys's goals for some combination of inputs, TechSys could use the model to investigate investment and management strategies with the existing architecture. If not, the model could be used to investigate the effect of modifying the architecture to achieve greater growth potential.

6.2.4.6 Identify the Applicable Views from the Enterprise Architecture Framework

The next step in the process of developing the simulation of TechSys's enterprise architecture was to formally identify the areas of the enterprise architecture that have an effect on the enterprise's ability to pursue and capture new business. The NREAF identifies eight views of the enterprise, but not all views are equally represented, and some, such as the services view, may not be represented at all in the simulation model.

After an initial analysis, the primary EA views driving the dynamics related to the simulation's problem were the strategy, organization, and process views. Each of these views was then modeled using a simulation approach appropriate for the dynamics in that view. In addition, some components of the knowledge, technology, and external/policy views were incorporated into the model. Figure 6.4 shows the general structure of the hybrid model. After developing this high-level abstraction of the model, each of the variables associated with the inputs and outputs of each of these submodels was identified, and diagrams of the high-level structure of each of the submodels were developed to ensure logical consistency across the system.

6.2.4.7 Hybrid Model Development

Using Figure 6.4 to gain an understanding of the interaction of various NREAF views, the structure of the model itself was then designed to be able to answer the original model question regarding cooperation among the operating units. Accordingly, the enterprise was modeled from an operating unit perspective, composed of multiple operating units with their own strategies, incentives, and processes. These operating units pursued business opportunities in their respective markets and in other synergistic markets depending on the amount of the bid and proposal budget and the percentage of pursued opportunities that had to be cooperative (synergistic). Opportunities identified by the operating units would then be funneled through each operating unit's business development processes in a stochastic manner, dependent on model inputs with regard to funding, resources, and research and development progress.

Each operating unit agent in the hybrid model encapsulates its own copy of the process submodel and strategy submodel, as well as any variables associated with the knowledge view, the external/policy view, and the information technology (IT)

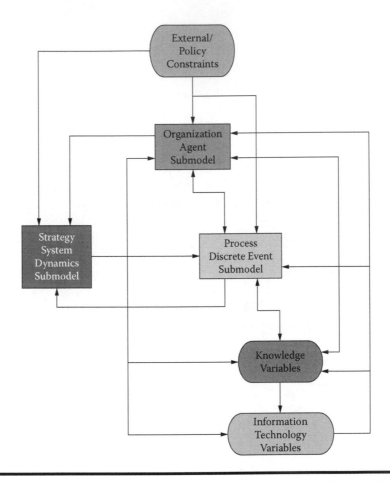

Figure 6.4 Logical interaction diagram of the TechSys enterprise architecture simulation model, based on the NREAF logical interaction diagram.

view. Figure 6.5 is a high-level "box" diagram of how a single operating unit would encapsulate these submodels and variables. When assembled into the full TechSys hybrid model, the structure appears as shown in Figure 6.6.

6.2.4.8 Simulation Software

The TechSys hybrid simulation model was implemented using the AnyLogic® software platform. AnyLogic is a flexible development platform for the development of deterministic or nondeterministic simulation models employing a wide array of simulation methodologies. This allows the creation of all three submodels within the same development environment, using the same timing and debugging engine,

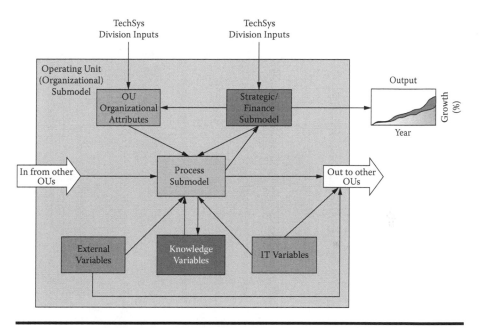

Figure 6.5 A high-level structural diagram of an operating unit with its encapsulated submodels and variables. (Reprinted from C.G. Glazner, *Journal of Transformation*, 1: 231–260, 2011. With permission.)

which greatly reduces the technical hurdles associated with creating hybrid simulation models. AnyLogic is a JAVA™-based application, and has the ability to export the model as a JAVA applet that can be run on any computer platform that has JAVA installed.

Following the object-oriented programming paradigm, AnyLogic allows the creation of models with multiple levels of encapsulation and inheritance, which makes the operating unit-centric topology possible. It also has tools for the development of graphical interfaces to the model that can be highly customized. Figure 6.7 highlights the initiatives of organizational elements leading to desired outcomes involving increased revenues, profits, or new products overlaid on the original model representation in AnyLogic. Every rectangle in the submodel view is an encapsulated model that can be viewed by selecting it. Parameters can be adjusted in a single table, and the initial value of variables can be adjusted by selecting it on the model chart.

6.2.4.9 Running the TechSys Simulation Model

The TechSys enterprise architecture simulation model is intended to be used as a tool to explore the potential performance of an enterprise architecture over a range of conditions. Each time the model is executed, the outcome will be unique, as the

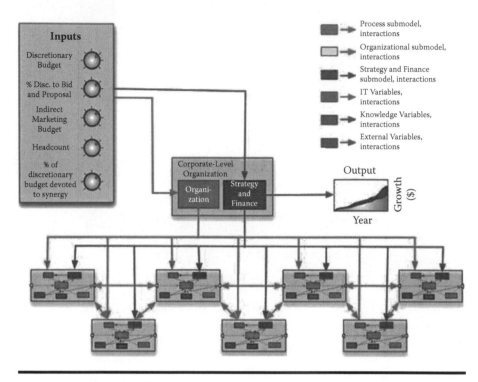

Figure 6.6   Top-level diagram, showing input and output parameters. (Reprinted from C.G. Glazner, *Journal of Transformation*, 1: 231–260, 2011. With permission.)

process submodel contains a number of nondeterministic steps. The key output is not the result of a single execution of the simulation, but rather the distribution of results from a number of executions of the simulation.

In keeping with the nomenclature of simulation modeling, a single execution of the simulation model using one set of input parameters is called a *replication*. Each replication will produce a unique result in a Monte Carlo fashion. A set of replications sharing the same input parameters is called a model *run*. Figure 6.8 is the performance distribution for a single run of the simulation with input parameters identical to those used in TechSys in 2007.

When comparing two different model configurations (either different inputs or different architectures), the probability distributions of their outputs should be compared, rather than the results of a single replication. Both the mean and variance of the output distributions should be compared. One model configuration, for example, might have a slightly higher mean than a second configuration but it may also have a higher variance. In this situation, the second configuration might be preferable because its lower variance would indicate lower risk and greater predictability.

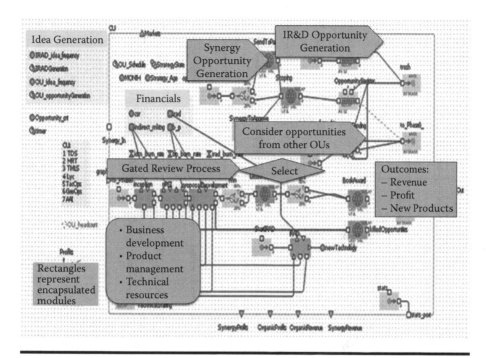

Figure 6.7 The operating unit submodel in the AnyLogic development environment. (Adapted from C.G. Glazner, *Journal of Transformation*, 1: 231–260, 2011. With permission.)

6.3 Analysis Using the TechSys Simulation Model

- At the highest levels of control, TechSys had a limited number of levers at its disposal to influence enterprise growth.
- The percentage of resources devoted to pursuing synergistic business opportunities.
- The discretionary budget.
- The amount of the discretionary budget allocated to bid and proposal versus internal research and development.
- Headcount.
- Indirect marketing budget.
- Changing the enterprise architecture itself.

For purposes of analysis using the simulation model, TechSys desired to keep the headcount and discretionary budget input parameters fixed, implying that no new resources would be used to foster new growth in the analysis. This left three options for influencing new enterprise growth in the model: the percentage of all new business opportunities that are synergistic, the percentage of the discretionary budget

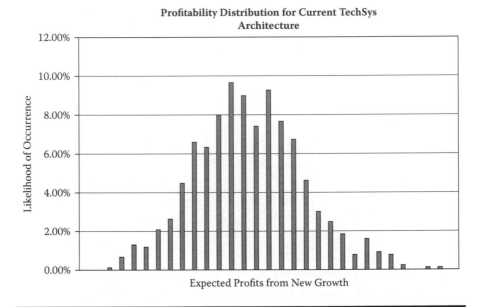

Figure 6.8 **The histogram representing the combined profitability from organic and synergistic profits for a single simulation run of the current state architecture. The independent axis values are withheld. *N* replications = 750. (Adapted from C.G. Glazner, *Journal of Transformation*, 1: 231–260, 2011. With permission.)**

allocated to bid and proposal (with the remainder going to internal research and development), and changing the architecture itself. The first step in the analysis was to test the current architecture "as is" to develop a baseline for any changes.

On the assumption that the current enterprise architecture was fixed, there were two key parameters that could be varied: the percentage of new synergistic business opportunities and the percentage of the discretionary budget allocated to bid and proposal. In the first analysis, the allocation of the discretionary budget was held constant, and only the percentage of new synergistic business opportunities was varied.

6.3.1 *Investment in Pursuing Synergy*

The expectation of the existing TechSys strategy is that as the percentage of synergistic business opportunities is increased relative to the amount of organic opportunities, the overall profitability of TechSys should also increase.* The reasoning for this is that synergy opportunities build the foundation for growth into new, more profitable markets. In theory, by pursuing synergy between its operating units, TechSys should have a stronger, more competitive position in the market. Over the preced-

---

* For a review of TechSys's usage of the terms "synergy" and "organic growth," see Section 6.2.1.

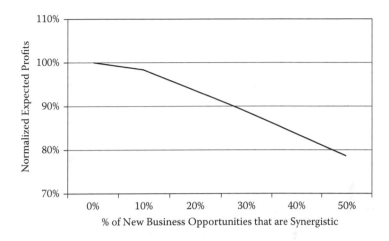

Figure 6.9   Expected profit as synergy investment is varied. (Reprinted from C.G. Glazner, *Journal of Transformation,* **1: 231–260, 2011. With permission.)**

ing four years, TechSys's synergy-driven strategy influenced many of the decisions that led to the development of the current state TechSys enterprise architecture. However, the output of the TechSys enterprise architecture simulation model tells a dramatically different story with regard to the benefits of pursuing synergy with the current state TechSys enterprise architecture.

Figure 6.9 indicates that in the TechSys enterprise architecture simulation, as the percentage of new business opportunities with a synergy component increases, enterprise profitability will decrease. This result strongly contradicts the prevailing theory of the effectiveness of pursuing synergy. On the surface, diminishing returns from pursuing a strategy based on synergy seems illogical. Upon examination of the behavior of the model, however, the reason for this significant inconsistency emerges directly from the enterprise architecture itself rather than from any shortcomings of effectiveness of pursuing synergy as an idea. The crux of the issue is that both the process and organizational architectures are structured such that synergy opportunities are systematically not selected compared to organic opportunities, leading to a large opportunity cost when more effective organic opportunities could have been pursued.

Perhaps one of the reasons that this shortcoming of the architecture had not drawn more attention previously is that it lies at the intersection of multiple views in the enterprise architecture framework. There are two key contributing factors that cause this process to favor organic opportunities: the first is the structure of the selection process itself, and the second is the organizational incentives of the operating units. These factors when combined have a multiplicative effect, causing the process's selection bias to be worse than analysis of the process and organizational incentives independently would suggest. The following sections present the problem from the perspective of each submodel, and then show how they interact to compound the problem.

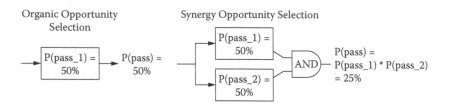

Figure 6.10 The mechanics of synergy and organic opportunity selection.

6.3.1.1 Process Submodel Perspective

In the model, all synergy opportunities are independently evaluated by each participating operating unit (OU) and prioritized against all other opportunities that each OU has. There is no central TechSys-level overview of synergy opportunities. For a synergy opportunity to be funded for further consideration, it must be independently chosen by all participating OUs during each OU's biannual opportunity review, where OUs pick which new business opportunities to fund in the coming months using funds from their bid and proposal and indirect marketing budgets. Because synergy opportunities must be independently approved twice by OUs with different local incentives, synergy opportunities are not selected at a rate exceeding new "organic" opportunities.

Figure 6.10 illustrates the mathematics of the opportunity approval process for both organic and synergy opportunities. On the left, an organic opportunity must only pass a single review. In the case of a synergy opportunity, it must pass two independent reviews. This has the effect of a Boolean logic AND gate. The resulting probability of the synergy opportunity receiving funding is equal to the product of pass rates from each OU's review, which can substantially lower the overall pass rate.

Because synergy opportunities are more likely to fail early in their development due to the selection problem above, devoting resources toward developing new synergy opportunities limits resources for developing good organic opportunities that are more likely to pass. This results in fewer high quality (in terms of profitability and competitive positioning) organic opportunities available for selection. With a smaller selection of organic opportunities to choose from, the expected value of opportunities that receive funding decreases.

6.3.1.2 Organizational Perspective

Taking an organizational perspective, the problem can be viewed in terms of local incentives for the OUs. The opportunity review selection team funds those business opportunities that are best poised to further the competitive position of the OU, based on a return to the OU of both profit and competitive position. On synergy opportunities, the profits from an opportunity are not evenly split between OU partners, nor is competitive positioning. Typically, there is a synergy partner that accrues the majority of benefit from pursuing the opportunity. The OU with the smaller share of profits and competitive positioning often finds it more advantageous to pursue its own locally developed organic opportunities, causing the synergy opportunity to go unfunded by both OUs.

The example in Figure 6.10 shows that each review has the same probability of passing an opportunity; in this example, P(pass) = 50%. This is not a realistic assumption, however. In practice, one OU will typically benefit more (either in terms of profit or competitive positioning) from a synergy opportunity than its partner will benefit, leading the total probability of a synergy opportunity passing to be lower than the OU with a lowest probability of passing the opportunity. In an example, if OU #1 has a 75% chance of passing a synergy opportunity and OU #2 has only a 25% chance of passing the same opportunity, the opportunity will have only an 18.75% chance of being funded.

This organizational behavior seems to occur despite the fact that the incentives of every general manager (GM) are aligned to promote the profitability of TechSys before that of the local OU. Interviews and discussions with TechSys stakeholders revealed two reasons why OUs still may not act in accordance with these incentives. First, the GMs have profit and loss responsibility for their OU. Despite their financial incentive plan, they tend to focus on the part of their compensation equation that they have the most impact on: their own OU. In TechSys's history, it has been exceptionally rare that OUs have worked together toward a common goal, providing little experience on whether and to what extent the GMs of the local OUs are willing to make decisions that might be detrimental at the business unit level while their actions help achieve global companywide goals. The second reason given that might help to explain the lack of effective incentives is that although the GM and some of the higher-level director positions may have a corporate- and division-level financial incentive scheme, the majority of people working on developing and selecting new business opportunities do not. Local incentives plus a lack of experience in working in a collaborative fashion lead to locally suboptimized behavior.

6.3.1.3 Additional Cultural and Communication Barriers to Synergy

In addition, synergy opportunities are more likely than organic opportunities to fail product life-cycle review gates even after they pass the initial opportunity review in the development pipeline. Communication difficulties and cultural alignment issues between OUs, included in the simulation model, cause the average failure rate for synergy opportunities to be higher than the average failure rate for organic opportunities, although the gap narrows over time as communication and cultural barriers are lowered due to repeated interactions between operating units (this forms an initial barrier to cooperation that will lessen over time).

The organizational submodel incentives compound the bias found in the process submodel for selecting synergy opportunities. Although the parallel selection requirement of the process significantly lowers the likelihood of a synergy opportunity being selected, the problem is exacerbated when there is a large differential in benefit and preference among the OUs, as is often the case in practice. Even if an individual synergy opportunity were vital to the strategy of one OU, it could be rejected because it would not benefit cooperating OUs as much as other local organic opportunities, despite the fact that TechSys as a whole would benefit more from the synergistic project. Given TechSys's strategic direction, this problem should be considered a critical area for improvement.

6.3.1.4 TechSys Corroboration

These results and the mechanism were corroborated with TechSys stakeholders. This behavior had been observed in practice, but synergistic business opportunities had remained such a small part of the total number of business opportunities considered to date that the systemic nature of the problem had not been highlighted. A working group at TechSys had identified the synergy opportunity selection process and incentives for OUs to pursue synergy opportunities as barriers, but had not realized the extent to which this barrier affects the system. This corroboration helps to confirm that this behavior is not purely an artifact of the modeling process, but is in fact present in the enterprise.

6.3.2 Allocating the Discretionary Budget

After understanding the effect of investing in pursuing synergy opportunities, the second major lever into its enterprise architecture that TechSys wished to investigate was the effect of the trade-off made when allocating the discretionary budget between the IRAD budget, and the bid and proposal budget. The discretionary budget is used by defense contractors to fund both IRAD, as well as bid and proposal activities. The discretionary budget must be divided between these two activities, creating a practical manifestation of the "exploration versus exploitation" trade-off common in the contingency theory literature. For many years, TechSys had allocated its discretionary budget without an explicit quantitative analysis of the impact of this allocation on their profitability. TechSys management clearly understood that investment in IRAD was necessary to ensure future growth, but also knew that without investing in pursuing new business using their existing capabilities, they would receive no new business. The simulation model can be used to better understand how this allocation decision affects the performance of the enterprise.

The TechSys enterprise architecture simulation model was used to analyze the effect of varying the allocation of the discretionary budget under the current state enterprise architecture. All model input parameters, other than "Discretionary

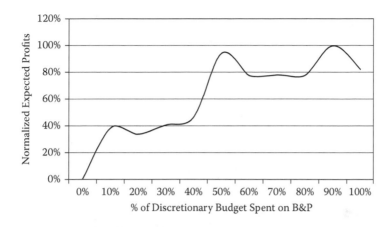

Figure 6.11 Normalized expected profit as the allocation of the discretionary budget is varied between IRAD (0%) and bid and proposal (100%). (Reprinted from C.G. Glazner, *Journal of Transformation,* **1: 231–260, 2011. With permission.)**

Budget allocated to Bid and Proposal," were held constant at their 2007 levels. The discretionary budget allocation to bid and proposal was varied from 0% to 100%, corresponding to the extreme cases when all discretionary money would be allocated to either IRAD or bid and proposal.

The result of varying the discretionary budget allocation in the simulation model is shown in Figure 6.11. The graph shows the expected profits over the three-year time horizon, normalized such that 100% is equal to the maximum value possible.

As can be seen in Figure 6.11, the simulation model indicates that given the current architecture, the preferred investment strategy is to allocate the majority of the discretionary budget to the bid and proposal budget, and less to the IRAD budget. When the majority of the discretionary budget is allocated to IRAD (for values less than 50% in Figure 6.11), the marginal benefit of allocating more money to bid and proposal is high; above 50%, the marginal benefit diminishes, but the graph still indicates that the preferred investment strategy would be to allocate somewhere between 50 to 90% of the discretionary budget toward bid and proposal. When the discretionary budget is entirely allocated to IRAD, the expected profit is $0, because no proposals have been written that would lead to a contract award and revenue. At the opposite extreme, if the entire discretionary budget were allocated to bid and proposal, there would still be a significant expected profit to be made, although there would be a slight dip from maximum profitability.

The trend shown in Figure 6.11 does not agree with common expectations for such a graph for an industry that produces many high technology products for the Department of Defense. Such a graph would be expected in a stable, commodity-focused enterprise where marketing and branding efforts have much greater impact than research and development. It is surprising then that this is a graph for an enterprise that produces advanced aerospace components, which then begs the

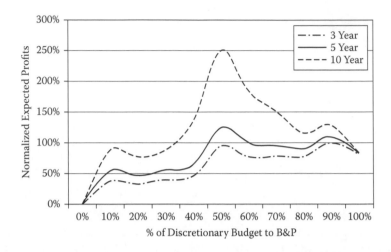

Figure 6.12 Expected profits as the discretionary budget allocation is varied, for time horizons of three, five, and ten years. (Reprinted from C.G. Glazner, *Journal of Transformation*, 1: 231–260, 2011. With permission.)

question of why the model is attributing such low impact to research and development at TechSys.

There are two factors at play in the simulation model that contribute to this behavior. The first is that the time horizon for the model is simply too short. The time horizon of the model was set to three years, because this is the stated strategic outlook of TechSys. That said, the benefits of research and development often take more than three years to show an impact on the profitability of the enterprise. Without extending the time horizon, the benefits of IRAD will not be apparent. An obvious test of this hypothesis is to rerun the simulation with a longer time horizon to see if the distribution of expected profitability changes to a great extent.

Figure 6.12 tests the effect of extending the time horizon from three years to five or ten years. As can be seen from the graph, moving from three years to five years emphasizes the peak at 50% of the discretionary budget allocated to bid and proposal, while slightly de-emphasizing higher allocations to bid and proposal. As the time horizon is moved to ten years, the effect is dramatically increased, and there is a clear preference for an even mix of investment between IRAD and bid and proposal. This figure indicates that TechSys should consider either extending the time horizon for the model, or keeping the time horizon at three years, but understanding that the long-term effects of IRAD are slightly undervalued.

A second explanation for the underperformance of IRAD is that IRADs pursued by TechSys do not have the impact that they should. Going back to the data gathered and the construction of the model, many of those interviewed during the data collection phase of the processes commented that TechSys takes a "peanut butter" approach to investing in IRAD projects: rather than choose areas for strategic investment that are tied back to strategy, the operating units tend to spend the money across many potential markets and technologies, without any particular focus. An internal study at TechSys in 2005 had trouble linking IRAD projects undertaken in the recent past with winning specific key proposals in that year, shedding further doubt on the effectiveness of IRAD investment at TechSys. As a result, the model randomly chooses technologies and levels of impact for each IRAD project, rather than selecting the best available from a pool of choices, as is done to select business opportunities. This diminishes the effectiveness of IRAD projects at securing a competitive advantage, especially compared to pursuing new business opportunities.

6.3.2.1 Discretionary Budget Allocation Dynamics

Figure 6.13 shows a simplified causal loop diagram that captures the primary dynamics resulting from discretionary budget allocation. As shown in the figure, both "Won Proposals" (through spending bid and proposal money) and "Technical Knowledge" (through spending IRAD money) can lead to increased business growth. As the competitive position increases, the probability of winning future proposals increases.[*] This is balanced by a "knowledge decay": over time, the potential for capturing new business decreases, in the absence of R&D spending, because the company finds itself exploiting its existing stock of knowledge rather than creating any new knowledge and hence technical advances through R&D investment. The connection between "Won Proposals" and "Competitive Position" is short term (approximately two years, depending on the knowledge decay of the market), as recent success can beget future awards. The connection between "Technical Knowledge" and "Competitive Position" is longer lasting (technical knowledge is not lost).

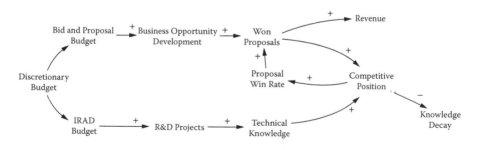

Figure 6.13 A simplified causal loop diagram of the dynamics of allocation of the discretionary budget. (Reprinted from C.G. Glazner, *Journal of Transformation*, 1: 231–260, 2011. With permission.)

---

[*] This relationship is one that is often observed in the industry; contract awards in one area increase the likelihood of future awards.

6.3.3 Combining the Levers: Performance Landscape for the Current State Enterprise Architecture

Thus far, the simulation model has been used to show the effect on expected profitability of varying two of the model's input parameters: the allocation of the discretionary budget and the percentage of new proposals that are synergistic. Each analysis was performed by varying a single parameter with all others held constant. Although this analysis has provided some insight into the performance characteristics of the enterprise architecture, it may be more useful to see how these parameters interact over the field of all possible combinations of inputs, as these inputs are not independent. Graphing the model's output as both parameters are varied results in a three-dimensional surface plot, with each input parameter shown on the *x*-axis and expected profitability shown on the *y*-axis. This surface can be thought of as a "performance landscape" for the enterprise, with peaks and valleys indicating the effect of different management strategies on enterprise performance. Figure 6.14 shows the performance landscape for the current state of TechSys's enterprise architecture when the input parameters "Percentage of synergy opportunities" and "Discretionary budget allocation to Bid and Proposal" are varied over the same ranges as previous analyses.

Looking at the performance landscape in Figure 6.14, the best possible expected profits can be achieved by not pursuing any synergy whatsoever and allocating a

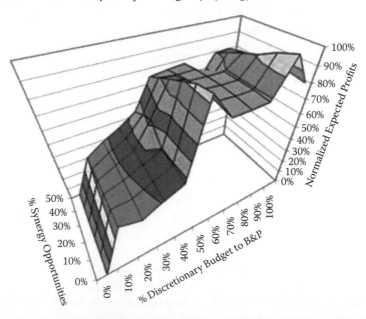

Figure 6.14 The performance landscape for the current-state TechSys enterprise architecture. (Reprinted from C.G. Glazner, *Journal of Transformation*, 1: 231–260, 2011. With permission.)

full 90% of the discretionary budget to bid and proposal, giving IRAD only 10%. The graph suggests the most robust strategy would be to take a position on the plateau that exists between 50 to 90% of the discretionary budget allocated to bid and proposal.

Many of the issues uncovered in the single variable analyses can also be seen in the performance landscape. As previously noted, as the percentage of synergy opportunities increases, profitability decreases in a monotonic manner, such that the maximum expected profitability occurs when there is no synergy whatsoever and each OU only pursues local, organic new business opportunities. As with the previous analysis of the discretionary budget allocation, the existing architecture favors a strategy where the majority of the budget is invested in bid and proposal, rather than in IRAD. Given the problems identified with the current enterprise architecture, it would not be difficult to imagine creating an alternative architecture that addresses these concerns, and is able to generate much higher expected profits.

6.3.4 *Creating an Alternative Architecture*

Working with TechSys stakeholders, an alternative enterprise architecture to the current state was developed to address the noted deficiencies of the current state architecture in the extremely limited sense pertaining to the allocation of available budgetary resources for capturing new business growth. The first changes to the architecture address the biases in the system against synergy. In the alternative architecture, synergy opportunities are not selected locally by the OUs, but rather by a team at the division level who select opportunities based on what has the most benefit for the division as a whole. The money for financing the new synergies will come by allocating a percentage of each OU's bid and proposal budget back to the division for synergy opportunities. The amount of this "synergy tax" on the OUs is the percentage of opportunities that have a synergy component (the same value as the model input parameter) multiplied by the OU's bid and proposal budget. The division chooses how this money will be spent, however, the OUs will still develop the ideas, and receive the benefits from winning the resulting contracts.

The second change to the current state architecture is to select IRAD projects in a more strategically aligned fashion. This change reflects changes that were underway at TechSys at the time the simulation model was completed. In the alternative architecture, IRADs are chosen based on their expected contribution to key technology areas that are aligned with a strategic technology roadmap. Although these roadmaps had been in use for years, they had not been used as part of the IRAD selection process.

The alternative architecture resolves the process bias, eliminating the independent approvals needed for synergy opportunities and not for organic opportunities, and also resolves the local OU incentives against choosing synergy opportunities.

The alternative architecture serves primarily as a "proof of concept" architecture, rather than as the blueprint for a future architecture that is under consideration. If the alternative architecture as described above were implemented, it would be met with widespread resistance from OUs, which stand to lose a substantial portion of their discretionary budget in the change. The alternative architecture is used here to develop a better understanding of the effects of changing the architecture, and as a starting point in future rearchitecting efforts.

6.3.5 Performance of the Alternative Enterprise Architecture

After coding the changes from the current architecture, the alternative architecture was run through the same evaluation runs as the current state architecture. Figure 6.15 shows the comparison of the current state architecture versus the alternative architecture on the dimension of investment in synergy. The data have been normalized so that 100% corresponds to the maximum expected profit under the current state architecture.

The first difference between the two curves is that for every point of comparison, the alternative architecture returns a greater expected profit than the current state architecture. More important, for many values, investing in synergy will produce an expected profit in excess of the maximum possible under the current architecture. Under the alternative architecture, synergy opportunities are able to have the impact that they were intended to have, increasing overall profitability by increasing competitive advantage and pursuing more profitable markets selling systems rather than components. At the peak of the alternative architecture's curve, the model indicates that a 22% increase in profits is possible by moving from the current state to the alternative architecture: a change that requires only a

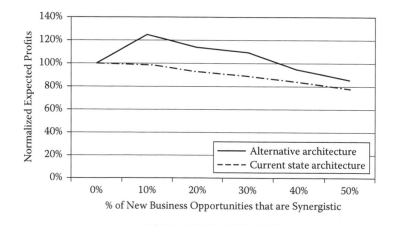

Figure 6.15 Expected profitability versus percentage of new business opportunities that are synergistic, for both the current state and alternative architectures. (Reprinted from C.G. Glazner, *Journal of Transformation*, 1: 231–260, 2011. With permission.)

small investment, a change in process, a change in incentives, and some amount of cultural consternation.

After peaking at a 10% synergy investment, the curve trends downwards again. Although it remains above the maximum possible under the current state architecture until an approximately 35% synergy investment, this downward trending behavior was not anticipated. Knockout analysis, where key elements of the structure were systematically removed between model runs, was used to determine the driving structure of this downward trending dynamic. This analysis revealed that this downward trend is attributable to the fact that there are a limited number of high-value synergy opportunities between the current operating units and once these high-value opportunities are exhausted, there is an opportunity cost associated with pursuing these opportunities rather than potentially more profitable organic opportunities. This misallocation of resources is due to the "walls" placed between the local OUs' bid and proposal budgets and the division's synergy budget, preventing a global "optimal" allocation of bid and proposal resources between organic and synergy opportunities across TechSys.

The relatively small number of quality synergy opportunities between OUs is attributable to the fact that there are not necessarily synergy opportunities between every pair of operating units. Of the seven OUs at TechSys, not all are in markets that could conceivably cooperate with others, causing these OUs to remain "blocked" from participating in synergy activities. In particular, one operating unit has little in common with the others, whereas two others have only limited potential opportunity for synergy. The potential for synergy collaboration between OUs is assessed in the model using qualitative, survey-based metrics, so there is a measure of uncertainty surrounding the exact number of potential synergy opportunities. The values used in the model should reflect a conservative estimate of synergy opportunities. As these qualitative values are changed, the location of the peak in Figure 6.15 moves. As synergy opportunities in the model increase, the peak moves to the right and increases in amplitude. To test out the model's sensitivity, the model is rerun with the qualitative parameter that measures synergy between each OU, which assumes a value ranging from 1 to 5 on a Likert scale, and is increased by 1 from its current value on the scale. As a result, the peak profitability outcome moves from its location at 10% up to 25% and increases in amplitude from 22% benefit over the base case to a 26% benefit.

Due to the sensitivity of the model to these qualitative parameters, the location and size of this peak must be evaluated with a measure of skepticism. Despite the uncertainty that exists in the graph, however, sensitivity analysis showed that the trend is robust to changes to model parameters. Even with the uncertainty as to the position and magnitude of the peak profitability level, the alternative architecture will have an appreciable increase in profitability associated with synergy investment with a peak, followed by diminishing returns.

Figure 6.16 shows the output of the alternative architecture while varying the allocation of the discretionary budget. The expected profit is shown for

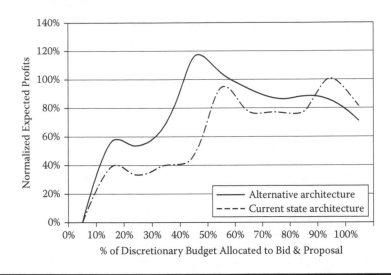

Figure 6.16 **Expected profits versus the percentage of the discretionary budget allocated to bid and proposal, for both the current state and alternative architectures. (Reprinted from C.G. Glazner,** *Journal of Transformation,* **1: 231–260, 2011. With permission.)**

both architectures, normalized such that the maximum possible from the current state architecture is 100%. As with the previous analysis of the allocation of the discretionary budget, the synergy investment parameter was held constant.

As can be seen in Figure 6.16, the alternative architecture favors a more balanced approach than the current state architecture. There is a clear preference in the model to allocate approximately 40% of the discretionary budget to bid and proposal, with the remainder going to IRAD. By increasing the effectiveness of IRAD through strategic selection rather than a "peanut butter" approach, simulation shows that an 18% increase in profitability can be expected. This change to the architecture also has the effect of skewing the curve to the left, giving more weight to the value of IRAD, as expected. As with previous analyses, this curve was produced using a three-year time horizon. When the time horizon is lengthened, the tail at the far right falls more quickly.

Figure 6.17 shows the enterprise performance landscape for the alternative architecture. As can be seen from the figure, the alternative architecture has a clear maxima that balances investments between bid and proposal and IRAD, and places value in a limited investment in synergy opportunities (10% of total development resources). There is no longer a flat stable region for higher allocations of the discretionary budget to bid and proposal, as seen in Figure 6.17; this architecture has a clear maximum, with a fairly steep decline in performance as more funds are allocated to IRAD.

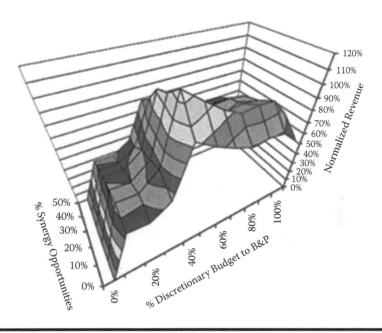

Figure 6.17 **The performance landscape for the alternative enterprise architecture.** (Reprinted from C.G. Glazner, *Journal of Transformation*, 1: 231–260, 2011. With permission.)

6.4 Recommendations from the Use of the TechSys Simulation Model

Based on this analysis, TechSys was able to identify key vulnerabilities in its organization and process design, as well as identify R&D expenditures as an area of concern. By addressing these issues, TechSys will remove barriers to successful execution on its strategies.

6.4.1 Benefits of a Hybrid Approach to Enterprise Architecture Simulation Modeling

Hybrid enterprise architecture simulation is a new approach to understanding enterprise dynamics, and the TechSys simulation model is the first of its kind to employ a hybrid simulation approach to analyze enterprise architecture from the multiple perspectives offered by enterprise architecture frameworks. The previous sections have shown how this approach has been successfully applied to address the pressing concerns of TechSys, however, it must also be noted that some of the problems observed at TechSys, such as synergy investment behavior, could only be addressed through a multiperspective approach. Hybrid simulation modeling is a useful tool in a modeler's toolbox, but for some classes of problems, it is the only tool.

There were two key dynamics at play in the TechSys simulation model: discretionary budget allocation and synergy investment. The first issue, discretionary budget allocation, can be fairly easily captured and modeled using system dynamics (assuming that data could be gathered that could capture aggregate characteristics of the processes). The causal loop diagram in Figure 6.13 can be expanded into a full-system dynamics model by quantifying the system in terms of stocks, rates, and variables, and the extra effort of interfacing multiple models would not be required. The issue of synergy investment, however, could not be so simply captured by any one modeling approach.

The dynamics of the synergy investment problem were much more complex than the discretionary budget problem. The problem lies at the intersection of two major views of the enterprise: process and organization. One component of the dynamics of synergy investment could be explained using a discrete event process model, similar to the one in the process submodel of the hybrid simulation model. This doesn't tell the whole story, however. The behaviors are also driven by the incentives of all of the operating units making locally rational decisions that end up producing suboptimal system-level outcomes. These dynamics were captured using the agent-based submodel in the hybrid simulation, which treated each OU as an agent with its own decision logic guiding its funding behavior. Without both the contributions of the process submodel as well as the organizational submodel, the full extent of the bias against synergy opportunities in TechSys's current state architecture would not be known.

Due to the flexibility and scope of the TechSys simulation model, both of the major parameters driving enterprise behavior could be varied simultaneously, providing the ability to create the performance landscapes shown in Figures 6.14 and 6.17. This would not have been possible without a hybrid simulation with an enterprise-level perspective.

One should not infer that hybrid modeling is the only approach that should be used to model the dynamics of enterprise architectures, but it does suggest that there is a class of problems that span the boundaries of architecture views that can only be fully addressed with this approach. Problems that can be described within the context of a single view can be modeled with a single simulation approach. Those that span views driven by very different behavioral dynamics (e.g., top-down versus bottom-up) may require a hybrid modeling approach to be applied. Without the aid of an enterprise framework to help with boundary setting and scoping, this class of problems has proven to be very difficult to detect and understand, as a mental exercise and from a simulation perspective. The application of enterprise architecture frameworks and hybrid simulation techniques to this class of problems provides an analytical approach that helps to manage the complexity of the dynamics and get to the causal structures and variables that drive enterprise behavior.

6.5 Conclusions

The TechSys Enterprise Architecture Simulation Model has proven to be a very useful tool to better understand TechSys's enterprise architecture and its effect on enterprise performance. Although the model does not make "crystal ball" predictions, it is capable of being used to understand how the enterprise architecture will tend to respond to varying the control levers into the architecture. This deeper understanding of the architecture can be used to think about how the enterprise can be managed and structured going ahead, and can be used as an input when making recommendations and decisions to increase the future performance of TechSys.

References

Abbott, A. 1990. Conceptions of time and events in social science methods: Causal and narrative approaches. *Hist. Methods*, 23: 140–150.

Dooley, K.J. and van de Ven, A. 1999. Explaining complex organizational dynamics. *Organization Science*, 10(3): 358–372.

Fowler, A. 2003. Systems modelling, simulation, and the dynamics of strategy. *Journal of Business Research*, 56(2): 135–144.

Glazner, C.G. 2011. Enterprise Transformation Using a Simulation of Enterprise Architecture, *Journal of Enterprise Transformation*, 1 (3): 231–260.

Hammer, M. and Champy, J. 1993. *Reengineering the Corporation: A Manifesto for Business Revolution*. New York: Harper Business.

Levitt, R. 2004. Computational Modeling of Organizations Comes of Age, *Computational & Mathematical Organization Theory*, 10: 127–145.

Mingers, J. and Gill, A. 1997. *Multimethodology: Towards a Theory and Practice of Combining Management Science Methodologies*. Chichester, UK: John Wiley & Sons.

Nightingale, D. and Rhodes, D.H. 2004. Enterprise systems architecting: Emerging art and science within engineering systems. Presented at the Engineering Systems Symposium. Cambridge, MA.

Rabelo, L. and Speller, T.H., Jr. 2005. Sustaining growth in the modern enterprise: A case study. *Journal of Engineering and Technology Management*, 22: 274–290.

Rabelo, L. et al. 2007. Value chain analysis using hybrid simulation and AHP. *International Journal of Production Economics*, 105: 536–547.

Rabelo, L., Helal, M., Jones, A., and Min, H. 2005. Enterprise simulation: A hybrid system approach. *International Journal of Computer Integrated Manufacturing*, 18(6): 498–508.

Schieritz, N. and Größler, A. 2003. Emergent structures in supply chains: A study integrating agent-based and system dynamics modeling. In *IEEE Computer Society: Proceedings of the 36th Hawaii International Conference on System Sciences*.

Simon, H.A. 1990. Prediction and prescription in systems modeling. *Operations Research*, 38(1): 7–14.

Sterman, J.D. 1991. A skeptic's guide to computer models. In G.O. Barney et al. (Eds.), *Managing a Nation: The Microcomputer Software Catalog*. Boulder, CO: Westview Press, pp. 209–229.

Sterman, J.D. 2000. *Business Dynamics: Systems Thinking and Modeling for a Complex World.* Boston, MA: McGraw-Hill.

Venkateswaran, J. and Son, Y.-J. 2005. Hybrid system dynamic—Discrete event simulation-based architecture for hierarchical production planning. *International Journal of Production Research,* 20(15): 4397–4429.

Womack, J and Jones, T. 1996. *Lean Thinking.* New York: Simon & Schuster.

Chapter 7

Reasoning on Technology Uncertainties for Enterprise Transformation

William J. Bunting[*]

Contents

[*] Adapted from Bunting, W.J. 2012. Reasoning on Uncertain Enterprise Technology Alignment for Insight into Attainment of Enterprise Transformation, *Journal of Enterprise Transformation*, 2(1): 50–79, January, 2012. With permission from Taylor & Francis Ltd, http://www.tandf.co.uk/journals.

7.1 Introduction

Enterprises comprise a multifaceted mixture of elements consisting of strategic outcomes, business processes, technology, and individuals. Strategic outcomes are the enterprise's planned accomplishments and represent what the enterprise values. Business processes consist of structures, data, and communication among company sites, roles, and divisions. Technology comprises the various systems that enable business processes. Individuals manage the complex interactions among these elements to produce outputs and attain a strategic outcome.

Enterprise transformation moves the enterprise from a given state to a new state in response to opportunities, perceived weaknesses, and changes in its domain (Kotnour 2011; Rouse 2005). It involves structuring the complex interaction under an organizational change management model, such as the Burke–Litwin model (Burke and Litwin 1992), which identifies systems as a key component in organizational change. Today, most large enterprise transformations require significant investments in new enterprisewide technology to accomplish the transformation (Merrifield, Calhoun, and Stevens 2008; Venkatraman 1994). This is especially true in the federal government (Government Accountability Office 2010).

This chapter does not discuss broad interactions of elements necessary to effect successful transformation. It instead focuses on an alignment reasoning method usable by managers involved in a technology-based enterprise transformation. To support a reasoned understanding of alignment, this chapter defines a technology initiative's enterprise line of sight (ELoS) as an inferential alignment of a technology initiative to business processes, business services, and strategic outcomes and supporting a principled assertion of attainment of the strategic outcome. This is shown in Figure 7.1.

The line of sight evidential reasoning analysis (LSERA) method specifically focuses on a technology initiative's ELoS that supports an enterprise transformation to attain some strategic outcomes. The LSERA method has two main uses:

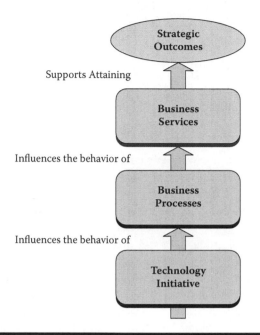

Figure 7.1 Enterprise line of sight. (Adapted from W. J. Bunting, *Journal of Enterprise Transformation*, 2(1, January): 50–79, 2012. With permission.)

- It supports modeling a specific technology initiative ELoS for examination (questioning, reasoning, inferring).
- It supports adjusting and then re-examining the representation on the basis of emergent information.

LSERA allows managers to reason about and gain insight from complex people, process, and technology interactions where enterprise analysts cannot easily construct algorithmic-based models.

7.1.1 Modeling Challenge

For a large complex technology initiative to support an enterprise transformation effectively, there must be a clear alignment of key aspects of the technology, the business processes directly affected, and the indirect effects on the business services composed of the business processes.

Managers must structure the alignment in a principled manner to assert with confidence that the technology initiative supports the strategic outcomes of the enterprise transformation. The challenges here are to identify: (a) the elements in the ELoS; (b) the states things can be in, such as attaining or not attaining; (c) what relates to what among technology, process, services, and information; (d) usability of data for reasoning; (e) the causal relations; and (f) a logic underlying the reasoning. In addition, managers must account for uncertainty in the ELoS. The challenges are to: (a) express uncertainty within the logic; (b) capture the manager's subjective beliefs about relations; (c) combine uncertain influences, such as a subprocess and technology influence on a process; and (d) identify the influential information underlining the ELoS argument.

A basic principle of LSERA is that enterprise transformation occurs when individuals who understand the complex element interactions prescribe correct actions. No set of managers, however, can understand all of these complex interactions. This presents two challenges. First, the enterprise needs a way for managers to construct localized models independently and at different times. Second, the enterprise transformation managers need a way to integrate these localized models in a principled manner into an overall model allowing them to examine the overall influences of localized changes. Last, as the transformation proceeds, managers maintain confidence in the key assertions by discerning emerging information, adjusting the models accordingly, and examining the changed results. What looked promising yesterday may look less so today. Emergent information, such as new innovative ideas, technologies, and changes in the environment, may strengthen or weaken the ELoS argument.

7.1.2 Enterprise Architecture Approach

Enterprises increasingly use enterprise architecture methods to support technology-based transformations and to support managers' understanding of complex element interactions. Among the main frameworks used are The Open Group Architecture Framework (The Open Group Architecture Framework 2009) and the Zachman Framework (Zachman 2008). To understand the interactions in federal agency technology initiatives, the federal government has specified the Federal Enterprise Architecture Reference Models (FEA-RM 2007), the Federal Segment Architecture Method (Federal Segment Architecture Method 2008), and, specifically for the Department of Defense, the Department of Defense Architecture Framework (Department of Defense Architecture Framework 2009).

These methods represent technology, process, and service entities and the relations among them, but they only model ELoS alignment as a set of relations. They inadequately specify the following: (a) a technology investment's ELoS as an ordered structure of technology, process, service, and outcome entities; (b) the relations among the entities' performance behaviors; (c) the relevance of information and modeling analyses to support inferences of performance behavior attainment; (d) the reasoning on the likelihood of attaining the enterprise's various performance targets and transformation; and (e) an approach for re-examining the likeliness of behavior attainments within the entire ELoS given each piece of new information. The LSERA knowledge structure derives from an examination of the enterprise architecture frameworks and from goal-oriented requirements engineering techniques (Chung et al. 1999; Van Lamsweerde 2001; Yu and Mylopoulos 1998; Zowghi and Offen 1997). To address the enterprise architecture framework limitations and the ELoS modeling challenges, LSERA uses evidential reasoning (Schum 1994) with multientity Bayesian network (MEBN) theory (Laskey 2008). LSERA consists of two main models: the argument model, which structures the reasoning and relevant evidence into a chain of IF–THEN statements; and the inference model, which combines local Bayesian models into an overall probability model indicating the likelihood of attaining a behavior, goal, or transformation.

The LSERA method supports three extensions to enterprise architecture and requirements engineering and:

- Considers enterprise architecture artifacts and other information artifacts as entities. These may be compound combinations of enterprise architecture metamodel entities, but they have no explicit representation within the enterprise architecture metamodel.
- Defines evidence entities. Evidence entities are instances of artifacts and metamodel entities that are relevant to an assertion.
- Defines inferential relations with probabilistic support, which enables reasoning on the likelihood of attainment.

7.2 Discussion of Key Concepts

The following provides a brief introduction to evidential reasoning and multientity Bayesian networks, key to understanding the LSERA method.

7.2.1 Evidential Reasoning

Reasoning is defined as the "use of reason, especially to form conclusions, inferences, or judgments" (American Heritage Dictionary of the English Language, Fourth Edition 2000). There are various types of reasoning: deductive (Johnson-Laird 2009), inductive (Hacking 2001), abductive (Josephson and Josephson 1996), analogical (Juthe 2005), case-based (Aamodt 1994), probabilistic (Pearl 1988), and evidential (Schum 1994). In an ELoS argument, analysts do not use (a) deductive reasoning, because there are no stated premises or valid structures (e.g., syllogism), (b) inductive reasoning, because there are no repeated instances of the alignment on which to assert the plausibility of attainment, (c) abductive reasoning, because they do not start with performance evidence and try to infer the most likely explanation for the evidence, or (d) case-based reasoning, because there are no appropriate cases to study. The reasoning is not analogical because, in general, the alignment argument is unique.

The ELoS argument has uncertainty within it, and analysts could choose to use some form of probabilistic reasoning. Probabilistic reasoning is reasoning that uses a form of probability theory as the argument logic. The ELoS has evidence and an argument structure, and analysts try to use the evidence to support or challenge an assertion. Thus, analysts can use evidential reasoning. In particular, they can use some form of evidential reasoning with probabilistic support.

Evidential reasoning is structuring evidence in support of, or against, a defined hypothesis. In general, an evidential reasoning argument consists of four parts: hypotheses, evidence, relevance, and force. Hypotheses are conclusions in which the argument authors have an interest. Hypotheses may be binary (e.g., true or false) or multiple (e.g., high, medium, or low). Hypotheses should be mutually exclusive and exhaustive. For example, a set of hypotheses about raining could be {it rained today; it did not rain today}. Analysts collect data to make hypotheses more than just beliefs. Data become evidence if the analysts can show that the data are relevant to one of the hypotheses. Relevance is linking evidence (data) to an hypothesis using a chain of reasoning. Generalizations support this linking. Generalizations are relations commonly understood in the context of the argument. Last, force of evidence refers to the evidence that, to some degree, changes the individual's belief in an assertion. A piece of evidence can be relevant but have little force.

7.2.2 Multientity Bayesian Networks (MEBN)

Any specific ELoS argument consists of multiple entities, attributes, relations, and uncertainty. The argument authors can express these ELoS elements as a Bayesian probability network (Jensen 2001). Given evidence findings, they then can assess the probability of a target variable given the evidence on hand.

A particular Bayesian network is not, however, applicable to other situations. It is situation specific; that is, it is a specific agency technology initiative. The network pertains to the analysis of a specific set of entities with set variables and attributes. Analysts can only enter findings related to the specific problem. This is restrictive for ELoS analysis, which begins with multiple types of different entities, the number of which may not be known beforehand (e.g., performance measures) and may have different relations. Defining a set of situation-specific Bayesian networks to analyze all of the ELoS arguments of interest may not be possible.

MEBN theory extends Bayesian networks by combining the expressive power of first-order logic with Bayesian probability theory (Laskey 2008). It represents a domain as a set of entities with attributes related to other entities. MEBN theory represents these relations as Bayesian network fragments, referred to as multientity fragments (MFrags). An MFrag consists of the following:

- Resident random variables that represent the focus of the MFrag
- Input random variables that are resident random variables from other MFrags that influence the MFrag local probability distributions
- Context nodes that are Boolean random variables representing conditions that must be satisfied
- A local probability distribution representing the uncertainty within the MFrag relations
- A fragment graph that expresses the conditional independence relations among the entities as a directed acyclic graph

MEBN represents specific unique instances with a unique identifier of the random variables. Given instances of the input random variables, an MFrag produces conditional probabilities and can result in a simple Bayesian network.

7.3 LSERA Method

This section describes the LSERA method. These are two main parts of the LSERA method: the argument model and the inference model. The argument model consists of six fundamentals: entities, hypotheses, directed relations, evidence, cause and effect, and logic. The inference model consists of four fundamentals: MFrags, probabilities, combination rule, and evidence force. Each of these fundamentals relates to one of the ten ELoS modeling challenges, shown in Table 7.1. The left column of Table 7.1 contains the modeling challenges, and the right column shows the corresponding LSERA fundamental.

In general, enterprise architecture frameworks contain the LSERA entities. A noted exception is evidence type and evidence instance entities. The frameworks do not contain hypotheses about the entities, although the frameworks can be extended to include hypotheses. The frameworks only have mappings between entities and not directed relations. They do not explicitly include causality or logic, and they do not support inference expression. Last, the frameworks have no structures to incorporate probabilistic relations. Thus, it takes considerable additional work to extract framework information and develop a situation-specific probability model.

Figure 7.2 is an overview of the general steps within the LSERA method. It starts with the technology initiatives manager expressing in natural language the technology initiative's enterprise line of sight. In Step 2, the analysts structure the ELoS by identifying the entities involved and their relations, evidence types, and causal influences. They then validate the logic of the argument model. Once the argument model is set, analysts move to Step 3, which starts the construction of the

Table 7.1 LSERA Fundamentals

Modeling Challenge	LSERA Fundamental
Argument Model	
1. The elements within the ELoS	Entities: Things or concepts of interest within the enterprise's domain
2. The states things can be in	Hypotheses: The states the entities may be in
3. What relates to what	Directed relation: A relation that indicates the direction of influence, A → B
4. Usability of data for reasoning	Evidence: Information that influences an individual's beliefs
5. The causal relations	Cause and effect: Criteria for a behavior to directly influence another behavior
6. Sound expression for reasoning	Logic: An IF–THEN expression, IF entity X is in state A, THEN entity Y will be in state B
Inference Model	
7. Expressing uncertainty within the ELoS	MFrags: MEBN fragments representing probabilistic relations within the ELoS logic
8. Beliefs of likely occurrence	Probabilities: A subjective measure representing the argument author's existing beliefs
9. Combining behavioral influences	Combination rule: Combining the influences of multiple behavioral entities
10. Influence of specific information	Evidence force: A measure of the influence of evidence in insolation or combination

Source: W.J. Bunting, *Journal of Enterprise Transformation*, 2(1, January): 50–79, 2012. With permission.

Note: ELoS = Enterprise line of sight; LSERA = Line of sight evidential reasoning analysis; MEBN = Multientity Bayesian network; MFrags = Multientity fragments.

inference model. This construction involves the analysts selecting the LSERA inference model fundamentals that correspond with the argument model. The inference model is then generated and validated.

In Step 4, analysts add into the inference model evidence artifacts, such as performance measurements or enterprise architecture artifact quality assessments, and the technology initiative managers examine the resulting probabilities. In Step 5, certain probabilities are adjusted to reflect the managers' indications of the relative

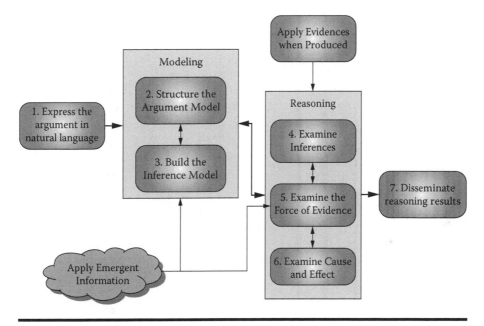

Figure 7.2 Line of sight evidential reasoning analysis method overview. (Adapted from W. J. Bunting, *Journal of Enterprise Transformation,* **2(1, January): 50–79, 2012. With permission.)**

forcefulness of specific evidence. In Step 6, managers examine the cause-and-effect relations to account for necessity and sufficiency conditions. In Step 7, managers disseminate the argument reasoning results to managers within the broader enterprise transformation for understanding and agreement. Later, as the technology initiative proceeds, new information emerges and is added into the argument and inference models, and the modeling and reasoning steps are executed as needed. This adjusting for new information continues until the technology initiative is completed.

7.3.1 LSERA Argument Model

The purpose of the LSERA argument model is to present the ELoS argument in a simple language and logic. This facilitates reasoning by enterprise managers and translating into a probability model by analysts. The argument model is a knowledge base consisting of six fundamentals: entities, hypotheses, directed relations, evidence credentials, cause-and-effect criteria, and IF–THEN reasoning. The following is a description of each of the six fundamentals.

7.3.1.1 LSERA Entities

LSERA entities are real things or concepts of interest in the enterprise's domain. For LSERA, the set of entities comes from a review of evidential reasoning theory,

MEBN theory, enterprise architecture frameworks, and the Federal Enterprise Architecture Performance Reference Model. These LSERA entities are what enterprise managers want to reason about when they assert that a particular technology initiative will directly or indirectly support the attainment of an enterprise's mission results, customer results, and strategic outcomes. There are 18 LSERA entities in four groupings: behavioral, performance, artifact, and evidential; see Table 7.2.

7.3.1.2 LSERA Hypothesis Sets

Given entities, managers can now make assertions about the entities, for example, that the process will attain its performance. An hypothesis set is how the enterprise expresses assertions about the LSERA entities. LSERA contains three hypothesis sets of mutually exclusive and collectively exhaustive statements. These are behavioral hypotheses, performance hypotheses, and information hypotheses. *Behavioral hypotheses* are how a real-world entity is likely to behave overall as a combination of its specific behaviors and in reaction to a specific set of conditions. The behavioral hypotheses are described as exceed, attain, and below relative to the desired behavior. *Performance hypotheses* are assertions about the measurable performance of the behavioral entities. The hypotheses are higher, within, and lower than the range of the specified target value. Enterprise subject matter experts define the acceptable range around the targeted value. *Information hypotheses* are assertions about the quality of supporting artifacts within the LSERA argument. The LSERA method defines quality broadly, and a particular enterprise will define it specifically for a particular argument context. The information hypotheses are high quality, acceptable, and low quality. For LSERA, these states indicate comparison of an artifact against an agreed-upon standard.

7.3.1.3 LSERA Directed Relations

A relation is a meaningful connection or association between two or more things on the basis of the relevance of one thing to another. If the relation indicates direction of influence, such as "*A* influences *B*," then this is a directed relation and it is represented by an arrow. A key characteristic of LSERA is a set of predefined directed relations. The LSERA argument model is not a freeform argument; it has a prescribed structure that forms a template for developing most LSERA ELoS arguments. This is different from enterprise architecture frameworks that permit defining any relation among their entity instances without expressing the nature of the relation.

LSERA has two kinds of directed relations: causal and inferential. In *causal*, the parent's (*A*) behavior influences the child's (*B*) behavior, as in, "*A* causes *B*." In *inferential*, the state of the parent simply infers something about the state of the child, as in, "Given an acceptable technical architecture, then it is likely the

Table 7.2   LSERA Entities

Behavioral Entities These entities are the actual services, processes, and technologies whose behaviors the enterprise wants to improve. Behaviors are externally observable characteristics of the entity that contribute to a result or strategic outcome. These five behavioral entities are represented within existing enterprise architecture (EA) frameworks and enterprise transformation theories.	
Mission Service	This is a specific service performed to attain defined mission and business results and customer results.
Processes and Activity	These are processes whose behaviors contribute to attaining mission service behaviors.
Subprocess and Activity	These are parts of a process and activity that are an important aspect of the overall process and activities behavior attainment.
Technology Initiative	This is the technology that directly contributes to attaining processes and activities behaviors.
Technology Initiative Aspect	These are aspects of a technology initiative whose behavior is critical to attainment of the technology initiative behavior.
Performance Entities. These are the measurable behaviors of the behavioral entities. Performance entities are used to show that the entity may attain its overall desired behavior. Performance entity constructs are found within existing EA frameworks and enterprise transformation theories.	
Strategic Outcome	This is the enterprise's defined strategic plan outcome that is the mission of the enterprise.
Mission and Business Result	These are measurable results focused on fulfilling specific areas of the enterprise's mission.
Customer Result	These are measurable results focused on supporting the enterprise's most external customers, such as citizens, businesses, or other enterprises.
Process Performance Measure	These are process performance Indicators measured to support reasoning on attainment of process and activities behaviors and subprocess and activity behaviors.
Technology Performance Measure	These are systems performance indicators defined to measure the emerging Technology Initiative behaviors and the related Technology Initiative Aspect behaviors.

continued

Table 7.2 (continued)   LSERA Entities

Artifact Entities. These artifacts support reasoning on a performance entity's targeted value, but they are not the actual measurement of a performance entity. The identification of specific artifact types that support reasoning on value attainment is a key distinction of LSERA.

Enterprise Architecture Artifact	These are types of artifacts, such as performance architecture, defined by and contained within an enterprise's EAs. Within LSERA the artifacts are looked at for their inferential force rather than simply their correctness. For example, a manager may infer that a weak technical architecture will decrease the likelihood of attaining a particular technical performance measure targeted value.
Other Information Artifact	These are any other types of artifacts that affect management's confidence in the attainment or nonattainment of desired results. They are found within the enterprise's knowledge domain. Examples are good standards, competitor product announcement, and law changes. Again, within LSERA the artifacts are looked at for their inferential force in addition to their primary use.

Evidential Entities. These are actual reported results that are instances of a performance measure entity or an artifact entity. The identification of artifacts as evidence within an argument is a key distinction of LSERA. LSERA defines evidence as information that influences an individual's beliefs regarding a particular assertion within an ELoS argument. Evidence is discussed in Section 7.3.1. EA frameworks may have performance measures and performance measurement entities but they do not have evidence as an entity or entity attribute.

EA Artifact Evidence	These are specific instances of EA artifacts selected for the technology initiative ELoS. Examples are versions of process models, technical architectures, and standards.
Other Information (OI) Artifact Evidence	These are specific instances of OI artifacts selected for the technology initiative ELoS. Examples are written engineering studies, new laws, and customer surveys.
Process Performance Measure Evidence	These are specific measurement observations done at a particular time that are related to a processes and activities performance measure.

Table 7.2 (continued)   LSERA Entities

Subprocess Performance Measure Evidence	These are specific measurement observations done at a particular time that are related to a subprocess and activities performance measure.
Technology Performance Measure Evidence	These are specific measurement observations made at a particular time related to a technology initiative performance measure.
Technology Aspect Performance Measure Evidence	These are specific measurement observations made at a particular time related to a technology initiative aspect performance measure.

Source: W.J. Bunting, *Journal of Enterprise Transformation*, 2(1, January): 50–79, 2012. With permission.

technology will attain its behavior." LSERA directed relations represent influence direction and associate the LSERA hypothesis sets with the LSERA entities. The following is an example of an LSERA directed relation statement: Technology Initiative X may likely *attain* its behavior, influencing process and activity Y to likely *attain* its behavior. Given the relevant hypothesis sets for any directed relation, the analysts can make nine inference statements for each directed relation. LSERA has 46 directed relations categorized into six groups. Figure 7.3 shows each of these groups graphically.

7.3.1.3.1 Group 1: Behavior-to-Behavior Directed Relations

These relations cover the five behavior entities. The general directed relation statement is: Behavior entity X may likely {Exceed, Attain, or be Below} its behavior, influencing Behavior entity Y to likely {Exceed, Attain, or be Below} its behavior. Within EA frameworks relations are defined to associate behavior entities, such as technology, with processes. However, they do not include the direction or influence inference statements.

7.3.1.3.2 Group 2: Behavior-to-Performance Directed Relations

These relations cover the five performance entities. The general directed relation statement is: Behavior Entity X may likely {Exceed, Attain, or be Below} its behavior, so the performance entity Y is likely to be {Higher, Within, or Lower} than its targeted range. This is an inferential extension of the relations found within the EA framework that simply relate performance to the behavior entity.

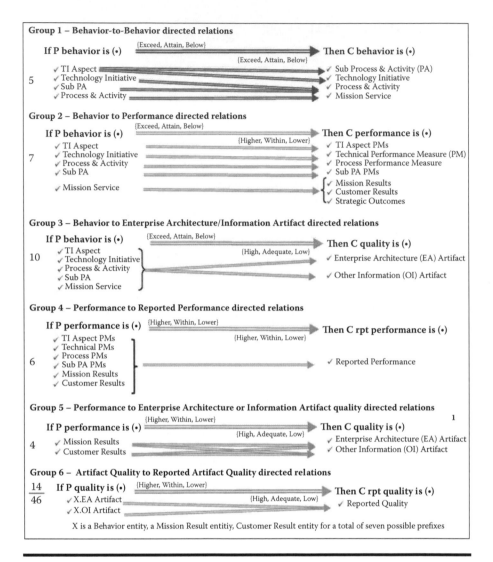

Figure 7.3 LSERA directed relations. (Adapted from W. J. Bunting, *Journal of Enterprise Transformation*, 2(1, January): 50–79, 2012. With permission.)

7.3.1.3.3 Group 3: Behavior-to-Enterprise Architecture/ Other Information Artifact Directed Relations

These relations relate the two artifact-type entities to the five behavioral entities, producing ten directed relations. The general directed relation statement is: Behavior entity X may likely {Exceed, Attain, or be Below} its behavior, so the artifact Y is likely to have {High, Adequate, or Low} quality. These relations are a key distinction of LSERA. For example, we are using an EA process model to support inferences on attaining the desired process behaviors and not simply documenting the process accurately.

7.3.1.3.4 Group 4: Performance-to-Reported Performance Directed Relations

These relations relate the technical and process evidence entities to the performance entities. The general directed relation statement is: Performance measure X entity may likely be {Higher, Within, or Lower} than its targeted range, so the reported performance measure evidence entity is likely to report {Higher, Within, or Lower} than the targeted range.

7.3.1.3.5 Group 5: Performance-to-Enterprise Architecture or Other Information Artifact Quality Directed Relations

These relations relate the artifact entities to mission results and business results entities. The general directed relation statement is: Performance entity X may likely be {Higher, Within, or Lower} than its targeted range, so the artifact entity Y is likely to have {High, Adequate, or Low} quality.

7.3.1.3.6 Group 6: Artifact Quality to Reported Artifact Quality Directed Relations

These relations relate the artifact entities to reported artifact quality prefixed by one of seven entities, namely the five behavior entities, a mission result entity, or a customer result entity. The general directed relation statement is: X.EA or X.OI artifact X may likely have {High, Adequate, or Low} quality, so the reported evidence Y is likely to report {High, Adequate, or Low} quality.

7.3.1.4 LSERA Evidence Credentials

LSERA defines evidence as information that influences an individual's beliefs regarding particular assertions with a specific LSERA ELoS argument. Within LSERA, there are the four evidence types: tangible, direct, ancillary, and emergent, and an LSERA ELoS argument may contain evidence that is contradictory, conflicting, corroborative, or convergent. For information to become evidence within an LSERA ELoS argument, it must have the following evidence credentials: relevance, credibility, and force.

- *Relevance.* For an item of information, new or existing, to have relevance within LSERA, four conditions must hold: (a) the analysts must map the information to one or more LSERA entities; (b) there must be a chain of reasoning (an ordered set of LSERA relations) connecting the information (evidence) to the LSERA entities in question; (c) each relation must be supported with some ancillary evidence; and (d) the information must involve one of the hypotheses within the entities. As analysts identify new information, its relevance needs to be determined using these four conditions before the analysts can add the information to a particular LSERA ELoS argument.

■ *Credibility.* This is the ability to inspire belief or trust. Understanding the credibility of information within an ELoS is fundamental to establishing the believability of the ELoS analysis. The LSERA ELoS argument is concerned with the credibility of the evidence of technical performance measure, process performance measure, enterprise architecture artifact, and other information artifact evidence. LSERA analysts assert credibility by attesting to the following two conditions: (a) *authenticity*, indicating that analysts have the evidence artifact under configuration control and can track changes made to the artifact; and (b) *accuracy*, indicating that the evidence artifact represents the technology initiative, associated processes, and mission services as established by standards.

■ *Force.* Evidential force is the amount and direction of influence. For information to become evidence, it must have force; the information must point toward an LSERA hypothesis and must change a manager's belief by an amount. Although it is good for managers to know the probability of an outcome, it is better if they understand which evidence is more forceful in making the outcome probable. Externalizing the manager's force belief allows the analysts to justify, question, and make inferences about the force and the associated evidence. The LSERA evidence force measurement is the Bayesian likelihood ratio. See the description of the LSERA inference model for an explanation of the Bayesian likelihood ratio.

7.3.1.5 LSERA Cause and Effect Criteria

The LSERA Group 1 behavior-to-behavior directed relations are causal; that is, one entity's behavior directly influences another entity's behavior. The *American Heritage Dictionary* defines *causality* as "The principle of or relationship between cause and effect." The "causality" has been a philosophical debate since Aristotle's *Posterior Analytics and Metaphysics* (1994). The modern debate can be said to start with David Hume's *A Treatise of Human Nature* (Hume 1739/2009). Today, there are many theories of causal inference (David 2004; Pearl 2000). For LSERA, a directed relation has cause and effect if it meets the following conditions: entity actuality; entity behaviors; entity pairs; temporal order; antisymmetry; proximity; and a causal level of positive relevance, necessary, sufficient, or necessary and sufficient.

Entity actuality means the enterprise managers have defined the behavioral entities within the directed relations as discrete things with specific boundaries. *Entity behaviors* can cause an effect that is externally observable and of interest to the managers. *Entity pairs* within each causal directed relation are modular and consist of parent–child pairs that are stable physical mechanisms where changes to a pair are local (i.e., they do not affect other pairs). *Temporal order* means that if Entity X behavior causes Entity Y behavior, then Entity X behavior must occur before Entity Y behavior within the same time interval. *Antisymmetry* means that if X causes Y, then Y will not cause X at the same time. Proximity means that Entity X behavior must have a single-step directed relation to the associated Entity Y behavior that does not allow inserting another step for the direct effect to occur.

Last, the *causal level* indicates the nature of the cause and effect among the behavioral entities. *Positive relevance* means that Y is more likely to occur given X occurs than if X did not occur. *Necessary* means that X is required to produce the desired effects in Y, and the desired effects will not be possible without X. *Sufficient* means that X's behavior alone very likely will produce the desired behaviors in Y, recognizing that there may be other ways to generate the Y behavior than with X. Necessary and sufficient means that X's behavior is necessary and X alone can produce the desired behavior in Y. Within LSERA, cause and effect is probabilistic. The analysts will need a probabilistic expression for positive relevance, necessary, sufficient, and necessary and sufficient to substantiate the direct or indirect cause assertions. The LSERA inference model supports this.

7.3.1.6 LSERA IF–THEN Reasoning

The LSERA basic reasoning construct is the IF–THEN statement. For LSERA, these statements have the form:

IF (Entity X is concluded as H_j) **THEN** (Entity Y may be concluded as H_k).

These statements are a refinement of the LSERA 414 inference statements, thus producing a set of 414 IF–THEN statements. For example, given the directed relation between the Technology Initiative entity and Process entity, argument authors can make this IF–THEN statement.

IF (Technology Initiative X is concluded as "Attain") **THEN** (Process Y may be concluded as "Attain").

The authors also can make eight other statements representing the different combinations of the hypotheses Exceed, Attain, and Below. Argument authors can combine these 414 statements into potentially thousands of arrangements to form the structure of a situation-specific argument. In addition, there are compound statements resulting from cumulative performance influences, conflicting performance influences, and behavior combination influences.

7.3.1.6.1 Cumulative Performance Influence

This is the effect of a number of performance entities (PE) with the same conclusion of H_1: Higher, H_2: Within, or H_3: Lower, on a hypothesis of a behavior entity (BE) with the hypothesis of H_1: Exceed, H_2: Attain, or H_3: Below. The logic is similar for evidence entities (Evd) with the hypotheses of H_1: Higher, H_2: Within, or H3: Lower, on a performance entity. The IF–THEN statement construct is:

> **IF** (PE_1 and PE_2 and ... PE_n are all concluded as H_i) **THEN** (BE may be concluded as H_k).

> **IF** (Evd_1 and Evd_2 and ... Evd_n are all concluded as H_i) **THEN** (PE may be concluded as H_k).

Within the federal government, this is the logic underlining the specification of the technology's key performance parameters. The Federal Enterprise Architecture (FEA) Performance Reference Model (PRM) and Department of Defense (DoD) standards express that there should be a manageable number of key performance parameters for any particular entity.

7.3.1.6.2 Dissonant Performance Influences

This is the effect of several performance entities where some are exceeding their targets, some are below their targets, and some are meeting their targets. The IF–THEN statement construct for two divergent performance entities is:

> **IF** (PE_n is concluded as H_i) AND (PE_m is concluded as H_j) **THEN** (BE may be concluded as H_k) where $n \neq m$ *and* $i \neq j$.

For multiple performance entities that are dissonant, the statement is

> **IF** ($\{PE_a\}$ are all concluded as H_1 AND $\{PE_b\}$ are all concluded as H_2 **AND** $\{PE_c\}$ are all concluded as H_3) **THEN** (BE may be concluded as H_k) where sets a, b, c are disjoint.

7.3.1.6.3 Behavioral Combination Influence

This is the effect of a number of behavior entities with the hypotheses of H_1: Exceed, H_2: Attain or H_3: Below, on a hypothesis of another behavior entity. The IF–THEN statement construct for cumulative influence is:

> **IF** (BE_1 and BE_2 and ... BE_n are all concluded as H_i) **THEN** (BE may be concluded as H_k).

The dissonant influence of a set of behavioral entities BE_n on another directly related behavioral entity BE is:

IF (({BE$_a$} are all concluded as H$_1$) **AND** ({BE$_b$} are all concluded as H$_2$) **AND** ({BE$_c$} are all concluded as H$_3$)) **THEN** (BE may be concluded as H$_k$) where sets a, b, c are disjoint.

LSERA expresses the logic of these cumulative and dissonant statements within the LSERA argument model. LSERA manages their probabilistic inference expressions within the LSERA inference model.

7.3.2 LSERA Inference Model

The purpose of the LSERA inference model is to support principled reasoning within ELoS arguments and examine the effect of emergent evidence on existing argument inferences. It does this by associating the argument model relations to Bayesian network fragments and then combining the fragments into a single model representing the ELoS argument. The inference model consists of MFrags, probability distributions, combination rule, and evidence force measurement.

7.3.2.1 LSERA MFrags

LSERA uses MEBN theory. MEBN theory supports encoding the knowledge structure of the LSERA argument model into Bayesian network fragments and combining the fragments into larger networks for inference and examination. The LSERA inference model is a set of 46 MEBN MFrags, one for each directed relation in LSERA. By combining the MFrags in a principled manner, the analyst produces a situation-specific Bayesian network representing the ELoS.

For example, the LSERA technology initiative performance measurement MFrag in Figure 7.4 presents the directed relation of the complex entry management system (ti) to Average Entry Request Response Time, one of its technical performance measures (tpm). This example is from the ELoS example in the next section.

This MFrag contains the following:

- A resident node that has the random variable performance (tpm). This random variable represents the performance entity and the tpm attribute represents a unique instance of a defined technical performance measure. The states of performance (tpm) represent the associated performance hypotheses.
- An input node that has the random variable behavior (ti). It represents the technology initiative entity and the ti attribute represents a unique instance of the TI. The states of behavior (ti) again correspond with the associated behavioral hypotheses.
- Three context nodes that define the existence of the technology initiatives, the performance measures, and the relations among the two within the specific ELoS argument.

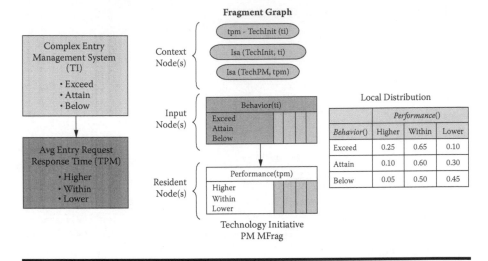

Figure 7.4 Sample multientity Bayesian network multientity fragment (MFrag); Avg = Average. (Adapted from W. J. Bunting, *Journal of Enterprise Transformation*, 2(1, January): 50–79, 2012. With permission.)

- A graph representing the directed relation.
- A local probability distribution that supports the IF–THEN logic of the associated nine likelihood statements and their probabilities.

This specific MFrag would be used for all of the other technical and process performance measures. With a few differences, all of the remaining LSERA MFrags are similar to this MFrag with a different local probability distribution and context nodes. The entire ELoS inference model is a simple principled combination of multiple instances of this structure. Once managers and analysts understand this simple structure, they can start to reason about alignment, usefulness of artifacts as evidence supporting assertions, and the probability of attainment.

7.3.2.2 Local Probability Distributions

The LSERA local probability distributions represent the "may be" expression within the directed relation. The distributions use a subjective probability measure representing the argument author's existing belief in the strength of the relation. For example, the statement IF (Technology Initiative X in state "attain") THEN (Process Y may be in state "attain") becomes the following statement on the basis of the argument author's expression of belief.

IF (Technology Initiative X in state "attain") **THEN** (Process Y in state "attain") is 70% probable.

	THEN *(Process Y in state " ")*		
IF (Technology Initiative X in state "Exceed")	Exceed	Attain	Below
	Is P_{11}	Is P_{12}	Is P_{13}

IF (Technology Initiative X in state "Attain")	Exceed	Attain	Below
	Is P_{21}	Is P_{22}	Is P_{23}

IF (Technology Initiative X in state "Below")	Exceed	Attain	Below
	Is P_{31}	Is P_{32}	Is P_{33}

Figure 7.5 IF–THEN probabilities structure. (Adapted from W. J. Bunting, *Journal of Enterprise Transformation*, 2(1, January): 50–79, 2012. With permission.)

The argument authors also can make eight other probability statements representing the different IF–THEN statements on the basis of combinations of the hypotheses *exceed, attain,* and *below.* Figure 7.5 shows all the combinations of this example in matrix form.

The probabilities must sum to 1 across the rows. By replacing the IF expression with the random variable *behaviorTI()* and the THEN expression with the random variable *behaviorPA(),* the structure is transformed into a conditional probability table as shown in Figure 7.6. When properly expressed, the probabilities p_{ij} make up the conditional probability table for the strength of the relation. Figure 7.7 shows the LSERA initial probabilities for the technology initiative to process directed relation.

There are 46 directed relations within LSERA that, given their respective random variables, produce 46 conditional probability tables containing 414 probabilities. The interaction of the probabilities is complex. If the conditional probability tables are not carefully specified, they can produce inconsistent results.

	behavior_PA()		
behavior_TI()	Exceed	Attain	Below
Exceed	P_{11}	P_{12}	P_{13}
Attain	P_{21}	P_{22}	P_{23}
Below	P_{31}	P_{32}	P_{33}

Figure 7.6 General probabilities. (Adapted from W. J. Bunting, *Journal of Enterprise Transformation*, 2(1, January): 50–79, 2012. With permission.)

	behavior_PA()		
behavior_TI()	Exceed	Attain	Below
Exceed	0.70	0.25	0.05
Attain	0.10	0.70	0.20
Below	0.05	0.25	0.70

Figure 7.7   LSERA behavior entity conditional probabilities. (Adapted from W. J. Bunting, *Journal of Enterprise Transformation*, 2(1, January): 50–79, 2012. With permission.)

Specifying each conditional probability table at the beginning of an ELoS argument development would be difficult for the analysts. LSERA has an initial set of six conditional probability tables, one for each directed relation set, and a prior probability table for the technology initiative aspect, as the only root node (Bunting 2009).

The fundamental probability elicitation technique is for the analysts to start with the LSERA initial probabilities and then adjust them on the basis of their review of the generated ELoS Bayesian network situation-specific inference model. This process consists of an ordered set of five to seven questions for each conditional probability table. The questions focus on specifying each of the nine likelihood probabilities, given harmonious and dissonant effects and behavior combination effects. This process assists enterprise argument authors in more principally expressing the probabilities within their specific inference model.

7.3.2.3 LSERA Combination Rule

With an LSERA argument, four situations require combining the influences of multiple behavioral entities. There are (a) multiple technology initiative aspects to a technology initiative, (b) multiple technology initiative aspects to a subprocess, (c) multiple technology initiatives to a single process and activity, and (d) multiple subprocesses to a process and activity. Each of these situations is identical within LSERA, and the Group 1 behavioral conditional probability table supports them. A combination rule is required, however, to describe the behavioral influences.

Using situation (b) as a means to explain the combination rule and a relation from the Entry Management Technology Initiative (EMTI) example from the ELoS example section, one question is how to model the combined influences of the two technology initiative aspects on the overall complex entry management system behavior. Figure 7.8 shows this case and the associated hypotheses. Note that these may not be the only technical aspects but are considered by the argument authors as the most influential in attaining the system's desired behavior.

Given the general unknown nature of influence combination within an ELoS, LSERA currently does not use independence of causal influence models such as

Figure 7.8 Example line of sight evidential reasoning analysis combination rule. (Adapted from W. J. Bunting, *Journal of Enterprise Transformation*, 2(1, January): 50–79, 2012. With permission.)

Noisy-MIN and Noisy-MAX (Diez and Druzdzel 2007). LSERA does use a combination rule called a *multiplexer rule* on the basis of Boutilier's multiplexer node (Boutilier et al. 1996). This multiplexer node addresses uncertainty in parent influence by establishing a random variable with a state for each of the parents, assigning a probability to each state that represents the parent's relative influence, and then constructing a conditional probability table with these additional state probabilities and the related parent and child conditional probability tables. Given this conditional probability table, a Bayesian algorithm determines the posterior probabilities of the child. If the relative contribution of the parent's influence to the total influences is unknown, the influence is distributed uniformly. Later argument modelers can adjust the influence probabilities according to new findings on the parent's relative influences.

For the example illustrated earlier, the multiplexer rule expresses that:

■ The admittance subsystem aspect may affect the influence of the identification subsystem aspect on the complex entry management system behavior and possibly vice versa, but the argument authors do not know how one aspect affects the other.

■ The argument authors assert that the total influence on the complex entry management system is probably coming from both subsystem aspects, but they cannot say which subsystem aspect is more influential, so they assign to each the same probability of influence, 50%, on the complex entry management system.

■ The total influence of both subsystem aspects is not synergistic; that is, it is not the sum of the two, nor is it detrimental, such as each possibly cancelling the other out.

As an example, using the LSERA probabilities, Figure 7.9 represents this influence situation. This multiplexer rule supports a quick look at influence combination. It is a starting point and, once the model is generated, the influence probabilities and the relationship probabilities can be changed to reflect influence combinations more accurately.

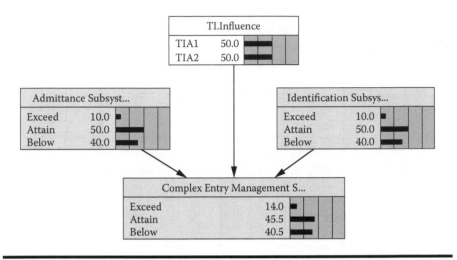

Figure 7.9 Sample use of multiplexer rule. (Adapted from W. J. Bunting, *Journal of Enterprise Transformation,* **2(1, January): 50–79, 2012. With permission.)**

7.3.2.4 LSERA Evidence Force

In an ELoS argument, enterprise personnel believe in the successful outcome of a technology initiative on the basis of particular evidence, but an ELoS argument is complex, and understanding the influences of, for example, contradictory evidence, is difficult. Calculating the force of evidence supports the manager's understanding of which pieces of evidence in isolation or in combination are contributing to the increase or decrease in the probability of a specific hypothesis being true. Managers understanding the force of particular evidence within the argument may increase their ability to make a more principled argument.

LSERA uses probability likelihood ratios to measure the force of individual items of evidence and the influence of evidence across the argument. The likelihood ratio is simply the ratio of the likelihood of a piece of evidence (e*) existing if something (call it H) occurred, divided by the likelihood of e* existing if H did not occur (Hc). Evidential force (shown as Le*) using the likelihood ratio is

$$L_{e*} = P(e*|H)/P(e*|Hc).$$

In the LSERA method, each entity has multiple hypotheses. For example, the behavior entity hypotheses are {*exceed, attain,* and *below*}. Thus, for LSERA, it is necessary to expand the likelihood ratio formula for multiple hypotheses, which is simply determining what is $P(e*|Hc)$.

Setting H_1 as *exceed*, H_2 as *attain*, and H_3 as *below*, and after some algebra, the evidential force of e* on, for example, H_2 is equal to its likelihood probability divided by the average of the likelihood probabilities of the other hypotheses. Equation (7.1) shows this force calculation.

$$L_{e^*} = \frac{P(e^*|H_2)}{\left[P(e^*|H_1) + P(e^*|H_3)\right]/2} \tag{7.1}$$

An assumption for force calculation is that all prior probabilities P(Hi) are equal. This assumption is necessary to make the evidentiary force calculation practical. The Bayesian likelihood force measurement is a ratio that has no upper bound but has zero as a lower bound. Thus,

- If $Le^* = 1$ then the evidence e^* does not favor the specific hypothesis over any of the other hypotheses.
- If $Le^* > 1$ then the evidence e^* does favor the specific hypothesis over the other hypotheses.
- If $Le^* < 1$ then the evidence e^* favors the other hypotheses over the specific hypothesis.

Because force is measured on an interval scale, there is no meaning behind the statement that a particular piece of evidence that has Le* = 4 is twice as forceful as another piece of evidence with Le* = 2 or, similarly, that Le* = .5 is twice as forceful in the opposite direction as Le* = 2.

The individual's interpretation of the nature of the probabilities within the context they are analyzing determines the significance attached to the magnitudes and differences of the force. This evidence force measurement is about the direct impact of a single piece of evidence on a single hypothesis. The calculations for other situations and combinations of evidence that may be harmonious or dissonant are more complex (Schum 1994).

7.4 ELoS Example

The LSERA method is presented using a federal technology initiative program. This federal technology initiative, for the purposes of this chapter, the EMTI, is part of the broad entry management services of goods and persons into the United States. It is a complex enterprise transformation involving daily interaction among a large number of federal agencies. The program had been in existence for more than five years at the time of the case study.

The purpose of the transformation is to eliminate redundant information requirements in support of efficiently regulating commerce and effectively enforcing laws and regulations. The primary enabler of the transformation is a new computer technology. The underlying technology will improve U.S. entry management processes, meeting two intermediate results of (a) increasing interdictions (mission result) and (b) facilitating entry into the country (customer result) while achieving

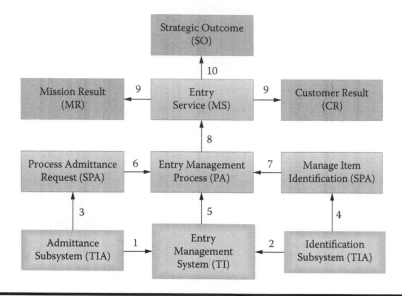

Figure 7.10 Entry Management Technology Initiative (EMTI) enterprise line of sight. (Adapted from W. J. Bunting, *Journal of Enterprise Transformation*, 2(1, January): 50–79, 2012. With permission.)

the overall strategic goal of preventing hazardous admissions into the country. The argument authors' main assertion is that if the agency implements the technology, then the agency will attain the aforementioned results and strategic goal. This is the technology's enterprise line of sight.

Figure 7.10 shows a simplified version of the EMTI argument. The arrows numbered 1 through 8 represent causal relations. The arrows numbered 9 and 10 are evidence relations. The numbers on the arrows relate to the inferences within the argument. The supporting inferences in the figure are as follows:

1. Technology staff asserts that if the admittance subsystem attains its desired behaviors, then the entry management system likely will attain its desired behaviors.
2. Technology staff asserts that if the item identification subsystem attains its desired behavior, then the entry management system likely will attain its behavior.
3. Technology staff asserts that if the admittance subsystem attains its desired behavior, then the subprocess process admittance request likely will attain its desired behavior.
4. Technology staff asserts that if the item identification subsystem attains its desired behaviors, then the manage item identification subprocess likely will attain its desired behaviors.
5. Technology staff asserts that if the entry management system attains its desired behaviors, then the entry management process likely will attain its desired behaviors.

6. Process staff asserts that if the process admittance request subprocess attains its desired behaviors, then the entry management process likely will attain its desired behaviors.
7. Process staff asserts that if the manage item identification subprocess attains its desired behaviors, then the entry management process likely will attain its desired behaviors.
8. Process staff asserts that if the entry management process attains its desired behaviors, then the entry service likely will attain its desired behavior.
9. Agency management asserts that if the mission service attains its behaviors, then the agency likely will attain its mission results and customer results. The agency's strategic planning process determines these results.
10. Agency management asserts that if the entry service attains its targeted behaviors, then the agency likely will attain its strategic outcome.

The key inferences are as follows:

■ If the entry management system attains its desired behaviors, then the strategic outcome of hazardous admissions prevention likely will be achieved to a significant degree.
■ If the entry management system attains its desired behaviors, then the agency goals (mission results and customer results) likely will be achieved to a significant degree.

In addition to these inferences, the process and technology staffs have defined several performance measures, enterprise architecture artifacts, and other information artifacts. These measures and artifacts support asserting whether the technology and processes will attain their desired behaviors. The staffs defined these measures to be independent of each other. This simple argument structure is the result of performing Step 1 of the LSERA method as shown in Figure 7.2. The following represents the EMTI argument model and inference model.

7.4.1 EMTI Argument Model

Step 2 in the LSERA method is the construction of the argument model. EMTI managers, analysts, and I structured the argument model over a period of several months. There was a large body of existing materials to start with, and the behavioral entities were easily agreed to. Determining the performance entities resulted in significant discussion. There were three reasons for this: determining the appropriateness of existing measures, the LSERA requirement that the performance entities be independent, and the requirement that performance entities must be formulated so that the "higher" state is good and the "lower" state is bad. We reached consensus on a set of performance entities and then selected the enterprise architecture artifacts entities. Program individuals were not familiar with enterprise architecture, but they came to agreement on the use of three types of technical architectures artifacts, four types of process models, an overall operational architecture type, and a performance architecture type. The directed relations among all of the entities were easily established. The next activity was asserting the cause and effect level. The managers had significant concerns about asserting strongly that the technology was necessary or sufficient. They preferred the default causal level of positive relevance. Last, the overall IF–THEN logic was examined and found acceptable.

Over several months the argument model evolved and concluded with the EMTI argument model graphic in Figure 7.11. The entire EMTI argument model includes 88 entities. The 88 entities are 3 technology and 4 process behavior entities, 18 performance measure entities, 1 strategic outcome entity, 1 mission result entity, 1 customer result entity, 9 enterprise architecture artifacts, and 12 other information artifacts. In addition, there are 39 evidence entities related to the performance measure, enterprise, architecture, and other information artifacts. Using the LSERA-directed relation set resulted in 89 relations; 8 causal shown in thick arrows and 81 evidential relations shown in thin arrows. To keep this graphic simple, it does not show individual relations of each of the evidence artifact entities.

7.4.2 EMTI Inference Model

Step 3 in the LSERA method, as shown in Figure 7.2 is constructing the EMTI inference model. The managers and analysts within the program had a limited background in probability theory. Thus, they used the default LSERA probability tables with little discussion. The LSERA MFrags that were selected corresponded to the argument model directed relations. Figure 7.12 shows a version of the EMTI inference model with initial evidence findings that all measures are within target and all artifacts are of adequate quality. The evidence nodes are hidden for readability.

Quiddity from OnLine Star, Inc., was used to join together the Bayesian network fragments under specific joining rules to generate the EMTI inference model. Quiddity passes the results to NETICA software for presentation. The EMTI inference model is a Bayesian network model consisting of 90 nodes and 91 relations.

At this point, managers and analysts can start the reasoning portion (Steps 4, 5, and 6) of the LSERA method, Figure 7.2, and discuss various situations of evidence combinations, evidence force review, and causal influence combinations. They can externalize their subjective beliefs about what is likely to happen and what they infer from new evidence. They do this by adjusting the probabilities within each relation. In addition, the argument authors, being experts in only their specific areas and possibly not in the entire argument, can do their analysis separately, with their thinking combined into a collective whole in a principled manner. With this inference model, the managers and analysts can now reason about the original EMTI inferences in a more principled manner. They can see what likely affects what and can reason about the implications of harmonious, dissonant, and

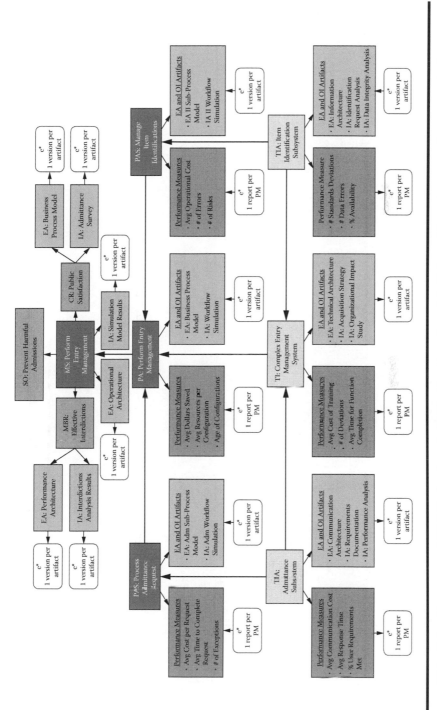

Figure 7.11 Entry Management (EM) Technology Initiative Argument model. (Adapted from W. J. Bunting, *Journal of Enterprise Transformation,* **2(1, January): 50–79, 2012. With permission.)**

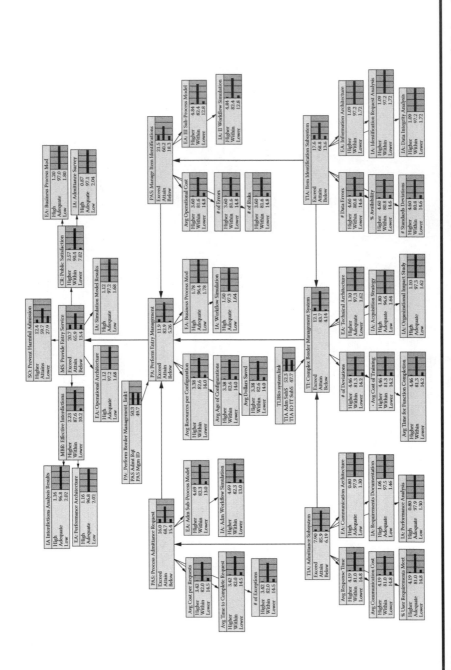

Figure 7.12 Entry Management Technology Initiative Inference Model. (Adapted from W. J. Bunting, *Journal of Enterprise Transformation*, 2(1, January): 50–79, 2012. With permission.)

new information. In addition, they can calculate the force of individual pieces of evidence to see which piece of evidence is driving the argument.

For example, a situation that was analyzed had the admittance subsystem significantly below its desired behavior. The identification subsystem's evidence indicated that it will likely attain its desired behavior. The complex entry management systems' performance measures were on target, and its enterprise architecture artifacts and other information artifacts were of good quality. The admittance subsystem substandard behavior and complex entry management systems' direct evidence resulted in the complex entry management system still having an approximately 70% probability of attaining or exceeding its desired behavior. The managers and analysts can now ask any number of principled questions on this result. One question to ask concerns the implication of the direct performance and artifact evidence versus the influence of the admittance subsystem behavior on the complex entry management system's behavior attainment probability.

Another question concerns whether these are the right performance measures to infer behavior of the complex entry management system. The managers and analysts can now determine if they feel that the model lacks accuracy and can adjust the performance entities, direct relations, and probabilities within these relations to reflect their collective beliefs.

7.5 Conclusion

The LSERA method provides the capability for managers to (a) develop a representation of a specific technology initiative within an enterprise transformation for examination (questioning, reasoning, inferring) and (b) re-examine the representation on the basis of emergent information.

LSERA provides the capability for enterprise analysts to develop an ELoS in a principled manner, extend it to all technology initiative aspects, and assist managers in reasoning on the probability of the technology's success given daily emergent information. Given software, the analysts can develop the LSERA models concurrently with the development of the ELoS argument. The time it takes to develop the two models is primarily based on the manager's ability to express the entities, relations, beliefs, and inferences. Analysts can quickly build the models and adjust them concurrently with changes in the managers' thinking. A key point to understand is that LSERA is most applicable in situations where it is not possible to build algorithmic models, that is, where reasoning and decisions are based on the beliefs of a group of managers such as in a large enterprise transformation. During the course of an enterprise transformation, enterprise architects, managers, and analysts develop numerous enterprise models and other analyses. Each result is relevant to assertions on attaining a desired aspect of the transformation. In general, the results are presented to a group of managers who then assimilate what they have heard, think about it at that moment, decide, and initiate actions. The problem is how to bring the results together in a principled manner to allow their combined influences to be examined. LSERA supports enterprise architects and other analysts in joining the information (evidence) into a principled model and permitting reasoned discussion on the results by management. The managers can then make informed enterprise transformation course adjustments.

References

Aamodt, A. 1994. *Case-Based Reasoning: Foundational Issues, Methodological Variations, and System Approaches*. Fairfax, VA: IOS Press.

American Heritage Dictionary of the English Language, Fourth Edition. 2000. Boston: Houghton Mifflin.

Aristotle. 1994. *The Complete Works of Aristotle: Revised Oxford Translation*, J. Boras (Ed.). Princeton, NJ: Princeton University Press.

Boutilier, C., Friedman, N., Goldszmidt, M., and Koller, D. 1996. Context-specific independence in Bayesian networks, uncertainty in artificial intelligence. In *Proceedings of the Twelfth Conference*, San Mateo, CA: Morgan Kaufmann.

Bunting, W.J. 2009. Reasoning on agency performance using line of sight evidential reasoning analysis. Paper presented at the *Uncertainty in Artificial Intelligence UAI2009 Seventh Annual Workshop on Bayes Applications*. Montreal, Canada (June).

Bunting, W.J. 2012. Reasoning on uncertain enterprise technology alignment for insight into attainment of enterprise transformation, *Journal of Enterprise Transformation*, 2(1, January): 50–79.

Burke, W.W. and Litwin, G.H. 1992. A causal model of organizational performance and change. *Journal of Management*, 18: 523–545.

Chung, L., Nixon, B.A., Yu, E., and Mylopoulos, J. 1999. *Non-Functional Requirements in Software Engineering*. Boston: Kluwer Academic.

David, A.P. 2004. Probability, causality and the empirical world: A Bayes–de Finetti–Popper–Borel synthesis. *Statistical Science*, 19: 44–57.

Department of Defense Architecture Framework. 2009. DoD Architecture Framework V2.0, Volume 2: Architectural Data and Models Architect's Guide.

Diez, F.J. and Druzdzel, M.J. 2007. Canonical probabilistic models for knowledge engineering (Technical Report CISIAD-06-01). Madrid, Spain: UNED.

FEA–RM (Federal Enterprise Architecture Records Management). 2007. Federal Enterprise Architecture Program Management Office, Federal Enterprise Architecture Consolidated Reference Model Document Version 2.3.

Federal Segment Architecture Methodology. 2008. FSAM version 1.0.

Government Accountability Office. 2010. Organizational transformation, a framework for assessing and improving enterprise architecture management. GAO–10–8466, Washington, DC: U.S. Government Accountability Office.

Hacking, I. 2001. *An Introduction to Probability and Inductive Logic*. New York: Cambridge University Press.

Hume, D. 2009. *A Treatise of Human Nature— Volumes I and II*. Frederick, MD: Merchant Books. (Original work published 1739.)

Jensen, F.V. 2001. *Bayesian Networks and Decision Graphs*. New York: Springer-Verlag.

Johnson-Laird, P. 2009. Deductive reasoning. *WIREs Cognitive Science*, 1: 8–17.

Josephson, J.R. and Josephson, S.G. 1996. *Abductive Inference: Computation, Philosophy, Technology*. Cambridge, UK: Cambridge University Press.

Juthe, A. 2005. Argument by analogy, argumentation. *Argumentation*, 19(1): 1–27.

Kotnour, T. 2011. An emerging theory of enterprise transformation. *Journal of Enterprise Transformation*, 1: 48–70.

Laskey, K.B. 2008. MEBN: A language for first-order Bayesian knowledge bases. *Artificial Intelligence*, 172: 140–178.

Merrifield, R., Calhoun, J., and Stevens, D. 2008. The next revolution in productivity. *Harvard Business Review*, June. http://hbr.org/2008/06/the-next-revolution-in-productivity/ar/1

The Open Group Architecture Framework. 2009. Version 9, The Open Group. Boston, MA.

OnLine Star Inc. IET. 2007. Quiddity (Computer software). Retrieved from http://www.iet.webfactional.com/quiddity.html.

Pearl, J. 1988. *Probabilistic Reasoning in Intelligent Systems*. San Francisco: Morgan Kaufmann.

Pearl, J. 2000. *Causality: Models, Reasoning, and Inference*. Cambridge, UK: Cambridge University Press.

Rouse, W.B. 2005. A theory of enterprise transformation. *Systems Engineering*, 8: 279–295.

Schum, D.A. 1994. *Evidential Foundations of Probabilistic Reasoning*. Hoboken, NJ: Wiley.

Van Lamsweerde, A. 2001. Goal oriented requirements engineering. A guided tour. In *Proceedings RE'01, Fifth IEEE International Symposium on Requirements Engineering*. Toronto: IEEE, pp. 249–263.

Venkatraman, N. 1994. IT-enabled business transformation: From automation to business scope redefinition. *Sloan Management Review*, 35: 2.

Yu, E.S.K., and Mylopoulos, J. 1998. Why goal-oriented requirements engineering. In E. Dubois, A. L. Opdahl, and K. Pohl (Eds.), *Proceedings of the Fourth International Workshop on Requirements Engineering: Foundations of Software Quality*, (REFSQ '98) Pisa, Italy, June 8–9, pp. 15–22.

Zachman, J. 2008. *The Zachman Enterprise Framework*. Retrieved from http://zachmanframeworkassociates.com.

Zowghi, D. and Offen, R. 1997. A logical framework for modeling and reasoning about the evolution of requirements. Paper presented at the *Proceedings of the Third IEEE International Symposium on Requirements Engineering*, Annapolis, MD.

Chapter 8

Optimal Control and Differential Game Modeling of a Systems Engineering Process for Transformation

Leonard A. Wojcik and Kenneth C. Hoffman

Contents

8.1 Introduction*

Large-scale, complex systems engineering efforts involving multiple stakeholders often have been problematic, so there is keen interest in both government and private industry in understanding how to improve the systems engineering process for such systems. This chapter presents an approach to modeling the systems engineering (SE) process, originally inspired by the highly optimized tolerance (HOT) framework for understanding complexity in designed systems. Here, high-level, optimal control-theoretic and differential-game models of the systems engineering process represent in a simplified way the cost, schedule, and performance tradeoffs inherent in systems engineering. Although the modeling is in an exploratory research stage and is not predictive, it is suggested that high-level models can be used to communicate risks inherent in new and ongoing systems engineering programs, as well as the potential effects of management and governance mitigation approaches. Key areas for further exploration include high-level modeling for systems engineering program diagnostics and early-stage assessment.

SE in the context of a large complex government or private enterprise is extremely challenging and many instances of such complex SE efforts have been problematic (Bar-Yam 2003). To cite a single, well-documented SE example dating back to the 1990s, the U.S. Federal Aviation Administration's (FAA) multibillion-dollar Advanced Automation System (AAS) program to modernize the U.S. air traffic control system took place in a context of major technical and management challenges together with unclear requirements in key areas, and was only partially completed, with major delays and cost overruns, when the program ceased to exist as originally conceived (Boppana et al. 2006). Boppana et al. (2006) present system dynamics and HOT models of SE processes applied to the AAS program, and pose the question of whether the complexity of SE in a complex enterprise such as civil aviation can be adequately represented and studied with such models. Here, we extend the unifying framework of Hoffman et al. (2007) as the basis for modeling the overall SE process at a high level, and present a range of optimal control and differential game models to account for SE complexity factors including uncertainty,

---

* Adapted, in part, from Wojcik, L.A. and K.C. Hoffman, © 2007 IEEE. IEEE/SMC International Conference on System of Systems Engineering (SoSE).

external influences (comprised of policy, operational, economic, and technological, or POET, factors [Carlock, Decker, and Fenton 1999]) and stakeholder interactions. The work presented here brings together previous research results as well as previously unreported results.

The high-level optimal control models of this chapter represent cost, schedule, and performance trade-offs inherent in the SE process. The models are not "predictive" because the uncertainties associated with POET and stakeholder factors are simply too large. Quantitative validation of the models is likely to remain limited, especially where POET factors dominate. Rather, the models are intended for use in playing out different assumptions and strategies in SE programs when POET and stakeholder factors pervade the SE process. Ultimately, with further research, the models might form the basis for greater scientific understanding of complex SE and contribute to the discourse on the nature of human/technological systems in general (Arthur 2009; Alderson and Doyle 2010). The models may prove useful in several ways relevant to practical SE.

First, they can serve as communication and educational vehicles to help SE practitioners understand the possible program implications of different SE management and governance approaches. As described later, the models can display emergent behavior patterns that may not be obvious to many government and industry decision makers and SE practitioners, yet are observed in particular SE programs. The models can help a wide range of decision makers, SE practitioners, and students understand these behavior patterns and how to prepare for or respond to them better.

Second, the models can be used for SE program diagnostics. When problems occur in an SE program, the models might suggest possible problem sources based on available data and other program information. A retrospective example of such a diagnostic analysis for the FAA AAS program is presented in this chapter. To do this model-based diagnosis, the parameters of a model were first fit to the original program plan, based on available information, then to the actual program trajectory, as it played out with various POET factors. Model parameter differences across the two program scenarios (planned versus actual) yield diagnostic information. As with diagnostic tools for other complex systems such as human health, experts are needed to interpret and apply the diagnostic results in the context of their knowledge and experience. But models of the SE process have the potential to extend what is possible with human expertise alone and also to communicate diagnostics clearly.

Third, the models can be used for early-stage assessment of possible SE management and governance approaches. Even though the models are not predictive, they can suggest qualitative types of possible outcomes and behaviors of which SE managers should be aware. Modeling for early stage assessment can augment expert judgment by "playing out" different possible assumptions about the future.

8.2 The Framework[*]

We use a general form for dynamics of the SE process based on optimal control theory as follows:

$$\dot{X} = f(X(t), U(t), t) \tag{8.1}$$

The vector $X(t)$ is a state vector for the enterprise and the vector $U(t)$ represents control elements affecting enterprise evolution. The components of $X(t)$ depend on the specific SE process being modeled, but might include, for example, metrics of progress toward the capability being developed, or total expenditures made toward the system acquisition, or other metrics related to the enterprise that could affect the SE process (Hoffman et al. 2007). The dot above X on the left side of Equation (8.1) is the derivative with respect to time. The function f relates the current state $X(t)$, the control $U(t)$, and time t to the evolution of $X(t)$ in time. In general, the function f may include random as well as deterministic variables. Following optimal control theory (Bryson and Ho 1975), we postulate further the motivation from stakeholder i for the controls stakeholder i exercises is expressed by an objective function for stakeholder i:

$$J_i = \phi_i\left(X(t_f), t_f\right) - \int_{t_0}^{t_f} L_i\left(X(t), U(t), t\right) dt \tag{8.2}$$

In general, different stakeholders involved in an SE effort may have different motivations, hence different J_i. For example, an SE contractor might be motivated by profit, whereas a government program manager might be motivated by the difference between benefit and total cost of the SE effort.

On the right-hand side of Equation (8.2), the second term integrates a cost function L_i from a specified start time t_0 to an end time t_f. In an acquisition context, the second term represents total cost to acquire the systems. The end time t_f, which typically represents the time to complete system development, may be a control variable. The end time together with the time dependence inherent in L_i specify a high-level "schedule" for acquisition and potentially other aspects of the SE process. The first term in Equation (8.2) represents the performance value of the acquired system in enterprise operations, which may depend on the time taken to acquire the system. Thus, Equation (8.2) represents mathematically the "iron triangle" of cost, schedule, and performance in SE.

A further generalization of Equation (8.2) is discussed later in this chapter, recognizing that there may be multiple stages toward development of a system, which may be completed across a range of times. At the end of each stage, it may be possible to field some capability into enterprise operations and generate benefit from the

---

[*] Wojcik, L.A. and K.C. Hoffman. 2007. ©2007 IEEE. IEEE/SMC International Conference on System of Systems Engineering (SoSE), April 16–18. With permission.

(partially completed) system. In this generalization, t_f can be understood as a vector of times corresponding to different developments, $X(t_f)$ as a matrix of states across the vector of times, and the integral of the cost function in Equation (8.2) can be expanded to a sum of integrals of cost functions across the different developments.

The assumption behind "optimization" in the optimal control framework for a single modeled stakeholder is that SE decision makers make choices reflecting cost, schedule, and performance trade-offs. It is not that SE decision makers literally optimize their decisions, but optimization in the framework shows the kind of high-level behavior that could be expected to emerge from cost, schedule, and performance trade-offs in an SE program. Similarly, with multiple stakeholders, equilibrium represents a type of collective behavior that could be expected among a set of stakeholders, each making its own trade-offs. Another important factor represented in the framework is that SE decision makers take into account not only the immediate effects of their decisions but also the longer-term effects. SE is a "planned" effort over a period of time, and the motivational framework expressed by Equation (8.2) reflects the impacts of a decision taken at time t on overall benefit accumulated over the system lifetime.

With only one stakeholder modeled, the problem of finding the "best" control $U(t)$ is an optimal control problem. With multiple stakeholders, different stakeholders exercise different controls, and there is not necessarily a single optimum for all stakeholders. We represent the states of the different stakeholders with different components of $X(t)$ and controls of the different stakeholders with different components of $U(t)$. Stakeholder interactions are modeled with a differential game approach, in which equilibrium solutions represent potential attractors for collective stakeholder behavior. As discussed later in the chapter, such equilibrium solutions are not necessarily unique for a given game, suggesting a mathematical basis for a fundamental unpredictability limit in multistakeholder models with scalar state and control variables.

8.3 Optimal Control Models

We begin with a discussion of optimal control models in which $X(t)$ and $U(t)$ are both scalars, and the objective function J also is a scalar. This is the simplest type of model in the optimal control framework, corresponding to a single modeled stakeholder with a unitary control.

8.3.1 Linear Models[*]

In a linear model, the equation of state (1) reduces to:

$$\dot{X} = A \cdot X + B \cdot U + \Delta \tag{8.3}$$

[*] Wojcik, L.A. and K.C. Hoffman. 2007. ©2007 IEEE. IEEE/SMC International Conference on System of Systems Engineering (SoSE), April 16–18. With permission.

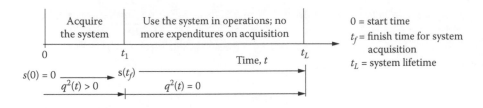

Figure 8.1 Base control-theoretic model scenario. (From Wojcik, L.A. and K.C. Hoffman. 2007. ©2007 IEEE. IEEE/SMC International Conference on System of Systems Engineering (SoSE), April 16–18. With permission.)

8.3.1.1 Objective Function Related to Operational Benefit [*]

We start with perhaps the simplest nontrivial instantiation of Equations (8.1) and (8.2). In this instantiation, the state X is a scalar function of time $s(t)$, the control U is a scalar function of time $q(t)$, the matrix B is a constant (denoted as b) and A and Δ are both zero. Figure 8.1 is a timeline for the generic SE scenario corresponding to this model, in which $s(t)$ measures progress made toward an operational capability in an acquisition.

8.3.1.1.1 The One-Third Rule[†]

The state $s(t)$ starts at zero at the beginning of the acquisition (time zero) and is unconstrained in the positive direction. The control function $q(t)$ is the positive square root of expenditure rate toward the acquisition. At time t_f the acquisition is assumed to be completed, at which time the acquisition expenditure rate goes to zero and the capability provided by the system is used in operations until a prespecified system lifetime t_L. Thus, transition to operations is incorporated in the acquisition period in this model. The acquisition completion time t_f is taken to be a control variable. The cost function L is $q^2(t)$ until t_f and zero afterwards. The operational benefit function φ is just the integral of the final state $s(t_f)$ from t_f to the system lifetime t_L, multiplied by a constant K, which corresponds to the dollar value of operational benefits over the period of use of the system in operations.[‡] Then, Equations (8.1) and (8.2) simplify to:

---

[*] Wojcik, L.A. and K.C. Hoffman. 2007. ©2007 IEEE. IEEE/SMC International Conference on System of Systems Engineering (SoSE), April 16–18. With permission. However, Section 8.3.1.1.4 is original to this chapter.

[†] Wojcik, L.A. and K.C. Hoffman. 2007. ©2007 IEEE. IEEE/SMC International Conference on System of Systems Engineering (SoSE), April 16–18. With permission.

[‡] We note that it is generally more realistic to assume decreasing benefits over the system lifetime, compounded by increased maintenance costs. These effects are treated in a later section of this chapter.

Table 8.1   Necessary Conditions for Optimal Controls

$\dfrac{\partial H}{\partial q} = 0$	Where: $\dot{s}(t) = f(s(t), q(t), t)$
$\dot{\lambda} = -\left(\dfrac{\partial H}{\partial s}\right)$	$J = \phi(s(t_f), t_f) - \displaystyle\int_0^{t_f} L(s(t), q(t), t) dt$
$\lambda(t_f) = \left(\dfrac{\partial \phi}{\partial s}\right)_{t=t_f}$	$H = f \cdot \lambda - L$
$\left(\dfrac{\partial \phi}{\partial t} + H\right)_{t=t_f} = 0$	$\lambda(t)$ is a Lagrange multiplier function.
$s(0) = 0$	

Source: Wojcik, L. A. and K.C. Hoffman, © 2007 IEEE. IEEE/SMC International Conference on System of Systems Engineering (SoSE).

$$\dot{s}(t) = b \cdot q(t) \tag{8.4}$$

$$J = K \cdot \int_{t_f}^{t_L} s(t_f) \cdot dt - \int_0^{t_f} q^2(t) \cdot dt = K \cdot s(t_f) \cdot (t_L - t_f) - \int_0^{t_f} q^2(t) \cdot dt \tag{8.5}$$

Equations (8.4) and (8.5) can be solved for the optimal controls $q(t)$ and t_f using the necessary conditions (Bryson and Ho 1975) in Table 8.1, which apply when boundary conditions are slack.

The optimization can be interpreted in at least two ways: first, prescriptively, it generates the expenditure trajectory and acquisition terminal time that produce the best result in terms of *J*. Second, borrowing from the HOT modeling approach for designed systems (Carlson and Doyle 1999), a descriptive interpretation of the optimization is that expert SE decision makers make decisions that result, at least approximately, in optimum overall performance. The second interpretation is the basis for comparing the results of the optimization to real SE programs.

The optimal control solution to Equations (8.4) and (8.5) is:

$$q(t) = \frac{K \cdot (t_L - t_f) \cdot b}{2} \tag{8.6}$$

for $0 \le t \le t_f$, and:

$$t_f = t_L / 3 \tag{8.7}$$

Thus, the optimum expenditure rate $q^2(t)$ is a constant that depends on the parameters K, b, and $t_L - t_f$. The state $s(t)$ is linear in time, and optimum acquisition time t_f is always one-third of total system lifetime, independent of the other parameters in the base model.

Taking the descriptive interpretation of the optimization, the model predicts that well-run, but complex SE efforts take approximately one-third of total system lifetime for acquisition including transition to operations, and two-thirds of total system lifetime for system use in operations. This "one-third rule" provides a benchmark for comparison with real SE programs.

8.3.1.1.2 Modeling "Internal Complexity"[*]

One seemingly arbitrary parameter in the base model is the square exponent in the cost term of J. If this exponent were unity instead (so that $q(t)$ is expenditure rate), the optimal solution would be "bang-bang"; that is, the optimum t_f is 0, and depending on the values of the parameters K and b, the optimum $q(t)$ is either zero or a delta function at time zero. We interpret the unit-exponent case as representing "zero internal complexity" in the acquisition, so that it is optimal to acquire as rapidly as possible. (We distinguish here between "internal complexity," which relates to the number and strength of interactions during the process of acquiring the system, and "external complexity," which relates to events external to the acquisition that affect operational requirements for the system being acquired or used in operations. External complexity is modeled later in this chapter.) For example, acquisition of a large number of identical and easily made parts would fit into the category of very low complexity: the acquisition can proceed extremely rapidly relative to the lifetime of the parts in the system because the means to manufacture them can be duplicated easily.

The case of a quadratic cost function discussed in the previous section represents a larger "internal complexity" for the acquisition. It is natural to generalize Equations (8.4) and (8.5) to include a variable exponent α in the cost function:

$$\dot{s}(t) = b \cdot q(t) \tag{8.8}$$

$$J = K \cdot s(t_f) \cdot (t_L - t_f) - \int_0^{t_f} q^\alpha(t) \cdot dt \tag{8.9}$$

Using the necessary conditions in Table 8.1, the optimal control solution to Equations (8.8) and (8.9) (where $\alpha > 1$) is:

---

[*] Wojcik, L.A. and K.C. Hoffman. 2007. ©2007 IEEE. IEEE/SMC International Conference on System of Systems Engineering (SoSE), April 16–18. With permission.

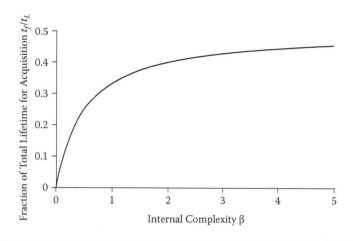

Figure 8.2   Fraction of total system lifetime for acquisition (From Wojcik, L.A. and K.C. Hoffman. 2007. ©2007 IEEE. IEEE/SMC International Conference on System of Systems Engineering (SoSE), April 16–18. With permission.)

$$q(t) = \left(\frac{K \cdot (t_L - t_f) \cdot b}{\alpha} \right)^{1/(\alpha-1)} \tag{8.10}$$

for $0 \leq t \leq t_f$, and:

$$t_f = t_L \cdot (\alpha - 1)/(2 \cdot \alpha - 1) \tag{8.11}$$

We interpret $\beta = \alpha - 1$ to be the "internal complexity" of the system, which has a range from zero to infinity. As β approaches infinity, t_f/t_L approaches a value of one-half. Thus, for SE programs with very high internal complexity, the model predicts that acquisition time will constitute about one-half of the total system lifetime. Figure 8.2 plots t_f/t_L as a function of internal complexity β. Note that for $\beta = 1$, t_f/t_L is 0.33, consistent with the "one-third rule" of the previous section.

8.3.1.1.3 System Aging Effects[*]

The base model assumes the operational value of the system is proportional to $s(t_f)$, throughout the period of system use from t_f through t_L. It is very likely that a real system will experience a decline in utility as it ages, and in addition, maintenance costs will increase with system age. There may be other time-dependent effects as well, such as a gradual introduction into operations, which might reduce system

---

[*] Wojcik, L.A. and K.C. Hoffman. 2007. ©2007 IEEE. IEEE/SMC International Conference on System of Systems Engineering (SoSE), April 16–18. With permission.

utility during the early part of its operational use. We model these effects by generalizing the objective function J to include a function of time $g(t)$ in the operational utility term, so the control problem is specified by:

$$\dot{s}(t) = b \cdot q(t) \tag{8.12}$$

$$J = s(t_f) \cdot g(t_f) - \int_0^{t_f} q^\alpha(t) \cdot dt \tag{8.13}$$

Solving Equations (8.12) and (8.13) for optimal controls using the necessary conditions in Table 8.1, we obtain:

$$q(t) = \left(\frac{g(t_f) \cdot b}{\alpha} \right)^{1/(\alpha-1)} \tag{8.14}$$

$$0 = \dot{g}(t_f) \cdot t_f + (1 - 1/\alpha) \cdot g(t_f) \tag{8.15}$$

For a simple case where the net benefit of a system decreases at a constant rate σ (taking into account both decreasing operational value and increasing maintenance costs),

$$g(t_f) = K \cdot \int_{t_f}^{t_L} (1 - \sigma \cdot t) \cdot dt = K \cdot \left[(t_L - t_f) - (\sigma/2) \cdot (t_L^2 - t_f^2) \right] \tag{8.16}$$

Substituting Equation (8.16) in Equation (8.15), and solving for t_f/t_L:

$$t_f/t_L = \frac{(2 \cdot \alpha - 1)}{(3 \cdot \alpha - 1) \cdot (\sigma \cdot t_L)} \cdot \left(1 - \sqrt{1 - \frac{2 \cdot \sigma \cdot t_L \cdot \alpha \cdot (3 \cdot \alpha - 1) \cdot \Omega}{(2 \cdot \alpha - 1)^2}} \right) \tag{8.17}$$

where:

$$\Omega = (1 - 1/\alpha) \cdot (1 - \sigma \cdot t_L/2) \tag{8.18}$$

Note that, in the limit as σ approaches zero,

$$t_f/t_L \to (\alpha - 1)/(2 \cdot \alpha - 1) \tag{8.19}$$

as expected from Equation (8.11).

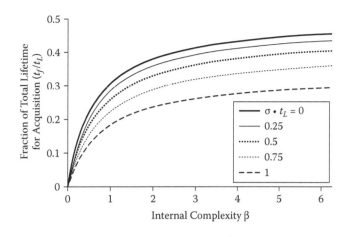

Figure 8.3 **Effect of internal complexity and aging on** t_f/t_L**. (From Wojcik, L.A. and K.C. Hoffman. 2007. ©2007 IEEE. IEEE/SMC International Conference on System of Systems Engineering (SoSE), April 16–18. With permission.)**

Figure 8.3 plots t_f/t_L in terms of internal complexity $\beta = \alpha - 1$ for several values of the product, $\sigma\, t_L$. Note that increasing σ reduces the optimum t_f/t_L.

8.3.1.1.4 Staged Development

The next extension of the scalar linear model is staged development, in which a partially completed system, with relatively limited capability, is introduced into operations, and additional system capabilities are generated in development. Here, $\beta = 1$ and no system aging effects are assumed. For a two-stage development, the objective function becomes:

$$J_{2\,stage} = K \cdot \left(s(t_{f_1}) \cdot (t_{f_2} - t_{f_1}) \ + s(t_{f_2}) \cdot (t_L - t_{f_2} - t_{f_1}) - \int_0^{t_{f_2}} q^2(t) \cdot dt \right) \quad (8.20)$$

Here, there are two "final" times, corresponding to the two stages. The first stage is completed at time t_{f_1} and the second at time t_{f_2}. It is assumed that, as soon as a stage is completed, it enters into operations and begins accruing benefit. This equation, together with the equation of state (8.4) can be solved as an optimal control problem, with controls $q(t)$, t_{f_1}, and t_{f_2}, using the equations in Table 8.1.

Analogous to the single-stage result presented before, optimal t_{f_1} and t_{f_2} are independent of the constants b and K:

$$t_{f_1}/t_L = 5/13$$

$$t_{f_2}/t_L = 23/39$$

$$(8.21)$$

And $q(t)$ is a step function:

$$q(t) = (4/13) \cdot b \cdot K \cdot t_L \text{ for } 0 \leq t < t_{f_1}$$

$$= (8/39) \cdot b \cdot K \cdot t_L \text{ for } t_{f_1} \leq t < t_{f_2} \tag{8.22}$$

$$= 0 \text{ for } t > t_{f_2}$$

A general result for an arbitrary number of stages has not been derived to date, but for infinite-stage development, we have:

$$J_{\infty \, stage} = \int_{t=0}^{t=t_f} K \cdot (t_L - t) \cdot ds(t) - \int_0^{t_f} q^2(t) \cdot dt \tag{8.23}$$

In this equation, t_f is interpreted as the time of the last infinitesimal stage. This optimal control problem can be solved in closed form, yielding the optimal $t_f = t_L$. The optimal values of J for one, two, and infinite steps compare as follows:

$$J_{1 \, stage} = \left(\frac{1}{27}\right) \cdot K^2 \cdot b^2 \cdot t_L^3 \approx (0.037) \cdot K^2 \cdot b^2 \cdot t_L^3$$

$$J_{2 \, stage} = \left(\frac{2672}{59319}\right) \cdot K^2 \cdot b^2 \cdot t_L^3 \approx (0.045) \cdot K^2 \cdot b^2 \cdot t_L^3 \tag{8.24}$$

$$J_{\infty \, stage} = \left(\frac{1}{12}\right) \cdot K^2 \cdot b^2 \cdot t_L^3 \approx (0.087) \cdot K^2 \cdot b^2 \cdot t_L^3$$

Note that J increases with the number of stages, which implies that development should include as many increments as possible to achieve maximum benefit. In real instances, engineering, development, and operational constraints and costs will limit the optimum number of stages. A practical SE implication of the multi-stage analysis is that the decision on how many stages to plan for can be framed as understanding the engineering, development, and operational limits to an infinite number of stages.

For the purpose of comparison with other models later in this chapter, the state trajectory with infinite stages is calculated:

$$s_{\infty}(t) = \left(\frac{b^2 \cdot K}{4}\right) \cdot (2 \cdot t_L - t) \cdot t \tag{8.25}$$

8.3.1.2 Objective Function Related to Pressure to Complete the System

In some cases, the modeler may wish to emphasize pressure to complete the system, rather than the benefit of the system in operations. For example, estimates of the operational benefit or the system lifetime might not be known. For these applications, a model with an objective function related to pressure to complete the system can be used. Pressure to complete the system might come from operational needs, or from political or organizational pressure, or a combination of both.

8.3.1.2.1 The "HOT" Model

HOT is a framework for understanding certain aspects of complexity in designed or engineered systems. Carlson and Doyle (1999) originally created the HOT concept and applied it to forest fire management, Internet traffic, and natural ecologies. For engineered systems, they showed how power laws relating event size to probability emerge from minimization of expected cost in the face of design trade-offs and uncertainty. Since then, the HOT methodology has been applied to such systems as the electric power grid (Alderson et al. 2004) and Internet architecture (Stubna and Fowler 2003). HOT typically is identified with power laws, but Wojcik (2004) applied a version of the HOT methodology to the SE process without finding power laws as a result; so it may be more correct to call it a HOT-inspired model, rather than a HOT model. In this section, we show that Wojcik's (2004) HOT-inspired model can be described in terms of the linear optimal control framework, but at the limit where the system is completed in an infinite number of incremental stages, as described above.

Wojcik (2004) posits a simple cost density function:

$$c(t) = A \cdot (1 - s(t)) \cdot (1 - e^{-t/\tau}) + B \cdot s(t) \cdot \delta_p(t) + D \cdot \dot{s}(t)^2 \tag{8.26}$$

Here, A, B, D, and τ are constants, and random variable $\delta_p(t)$ is a delta function with probability density p over time t. The parameter p is assumed to be a constant. The first term of the equation for $c(t)$ models the pressure to finish the base system, whether from actual system needs or other sources such as political pressure. The second term represents the cost incurred from random events that change how the system will be used relative to base system capabilities. This term incorporates stakeholder interactions as well as other external events. The second term is proportional to $s(t)$, which represents greater impact on a system that is closer to completion. The cost of actually building the system is modeled by the third term, which is nonlinear in $s(t)$, to reflect the relative difficulty of building the system in a short time compared to a longer time.

Furthermore, in this model $s(t)$ is constrained to be between 0 and 1. At a value of $s(t) = 1$, the system is completed and the pressure to complete reaches zero. Then, the expected cost over the lifetime of the system is:

$$\langle c_{tot} \rangle = \int_0^{t_L} \langle c(t) \rangle \, dt = \int_0^{t_c} L\big(\dot{s}(t), s(t), t\big) \, dt + \varphi\big(s(t_c), t_c\big) \tag{8.27}$$

where t_c is a cut-off time for development, whose default value is t_L, ($q(t) = 0$ for $t > t_c$). And,

$$L\big(\dot{s}(t), s(t), t\big) = A \cdot \big(1 - s(t)\big) \cdot (1 - e^{-t/\tau}) + B \cdot p \cdot s(t) + D \cdot \big(\dot{s}(t)\big)^2 \tag{8.28}$$

$$\varphi\big(s(t), t\big) = \int_t^{t_L} \Big\{ A \cdot \big(1 - s(t)\big) \cdot (1 - e^{-t^*/\tau}) + B \cdot p \cdot s(t) \Big\} dt^* \tag{8.29}$$

From the linear state equation (8.4) we substitute $b \cdot q(t)$ for $\dot{s}(t)$ in Equation (8.28), and obtain an objective function to be maximized:

$$J_H = -\int_{t_0}^{t_L} \Big\{ A \cdot (1 - e^{-t/\tau}) - M(t) \cdot s(t) + D \cdot b^2 \cdot q^2(t) \Big\} dt \tag{8.30}$$

where:

$$M(t) = A \cdot (1 - e^{-t/\tau}) - B \cdot p \tag{8.31}$$

For the case where the constraints are slack, this optimal control problem can be solved using the Hamiltonian approach expressed in Table 8.1. As originally reported by Wojcik (2004), the optimal state trajectory is:

$$s(t) = \frac{(A - B \cdot p)}{2 \cdot D}\big(t_L \cdot t - t^2/2\big) + \frac{A \cdot \tau^2}{2 \cdot D}\big(e^{-t_L/\tau} \cdot t/\tau - 1 + e^{-t/\tau}\big) \tag{8.32}$$

As τ approaches 0, the state trajectory $s(t)$ approaches the same trajectory as in Equation (8.25) for an infinite number of steps and objective function based on operational benefits, with:

$$b^2 \cdot K = (A - B \cdot p)/D \tag{8.33}$$

8.3.1.2.2 Application to the FAA AAS Program[*]

The FAA AAS program (1982–1994) is a well-documented example of a complex SE program whose result was considerably less than originally intended (Boppana et al. 2006). To apply the HOT-inspired model to the whole AAS program, Boppana et al. (2005) used the program plan and original budget to calibrate the values of *A, Bp, D,* and τ, as they appeared at the beginning of the AAS program. The theoretical basis for such a comparison is that there was one set of assumptions that was expressed (although not necessarily believed by anyone) in the original program plan, and another set of realities that actually played out in the execution of the SE program. The originally planned and actual spending profiles were used to generate two sets of HOT-inspired model parameters. It was assumed that competent SE decision makers faced with a very difficult and complex SE program tended to behave in ways motivated by the cost, schedule, and performance trade-offs reflected in the HOT-inspired model.[†] The control optimization step in the HOT-inspired model generates a program trajectory over time reflecting at a high level this kind of decision-making behavior.

No "original AAS plan" at a sufficient level of detail was found in the literature, so an approximation to the parameter set corresponding to an original plan was approximated from available information relating to expectations in the early 1980s about the AAS program spending profile. The parameter set derived for the original plan was: $A = 1$, $τ = 10$, $Bp = 0$, $D = 45$ (Boppana et al. 2005). This parameter set generates a steep spending profile from the initial year to a total of \$3.2B spent roughly 13 years after the initial year. This reflects an ambitious schedule that does not account for significant program uncertainties.

Then, a revised set of HOT-inspired model parameters A^*, B^*p^*, D^*, and $τ^*$ was estimated from the actual expenditure trajectory of the AAS program, using a design of experiment (DoE) based on an orthogonal array for four factors (the four HOT-inspired model parameters) at three levels and three confirmation runs. The actual AAS spending trajectory was based on congressional appropriations information documented in an AAS postmortem report by the Department of Transportation inspector general (1998). The HOT-inspired model parameter set corresponding to the actual expenditure trajectory is $A^* = 1.25$, $B^*p^* = 10$, $D^* = 60$, and $τ^* = 10$.

A comparison of the planned and actual AAS program expenditure trajectories is in Figure 8.4, which shows cumulative expenditure as a function of time.

---

[*] Adapted from K. Boppana et al., New England Complex Systems Institute (NECSI) International Conference on Complex Systems (ICCS2006). © 2006, The MITRE Corporation and Massachusetts Institute of Technology. All rights reserved.

[†] A conceptual limitation of the modeling is that the HOT-inspired model assumes the program is completed in infinite incremental steps, whereas the real program was not. At the time the study was done, the HOT-inspired model was the only suitable formation of the framework that was available.

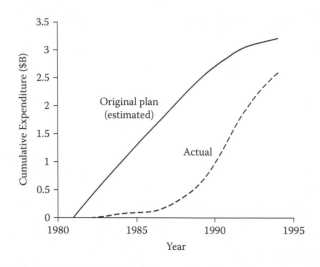

Figure 8.4   AAS planned and actual cumulative program expenditures.

A diagnostic interpretation is that the original AAS plan underestimated the level of uncertainty of the effort ($B^*p^* > Bp$), the inherent technical difficulty of the development ($D^* > D$), and external pressures on the program ($A^* > A$). From Figure 8.4, the effect of the difficulties (B^*p^*) immediately after contract award is visible as a slow startup. Then, later in the program, pressure to complete (A^*) forces up the expenditure rate, but the inherent difficulty (D^*) prevents the developed system capabilities from even coming close to the original plan.

Could the demise of the AAS program have been predicted at its start? Given the great uncertainties, a deterministic prediction would not have been useful, but a range of possible scenarios could have been played out with the HOT-inspired model. Information from expert panels reported by the U.S. Congress Office of Technology Assessment (1982) (before the name "AAS" came into common usage) suggests that some experts were suspicious of the program's technical viability, as well as there being some awareness of POET and stakeholder factors that were likely to increase the complexity of the SE effort. The range of opinions could have been played out in a set of HOT-inspired simulations to show a set of possible outcomes given the program strategy at the time. Alternative program strategies could have been assessed in terms of their robustness in the face of perceived uncertainties at the time, perhaps using extensions of the HOT-inspired model to generate high-level program trajectories, such as the expenditure trajectory of Figure 8.4. The utility of the models would have been to translate divergent expert opinion into a set of possible program trajectories. Broadly speaking, this approach is similar to model-based policy analysis to address such issues as sustainable development and pollution, where large uncertainties can dominate the analysis (Popper, Lempert, and Bankes 2005).

8.3.2 Nonlinear Model of External Complexity*

To model external complexity, in which events such as changes in enterprise mission or an evolving enterprise environment alter the requirements for a system, we modify the base control-theoretic model to allow for sudden "hits" to $s(t)$, which instantaneously reduce $s(t)$, and also "helps," which instantaneously increase $s(t)$. Hits and helps (which are called "hits" below for brevity) may occur during the acquisition period from 0 to t_f, or afterwards, during the period of operational use. For simplicity, the internal complexity parameter β was set equal to 1 and aging parameter σ was set equal to 0, as in the base model. The new equation of state is:

$$\dot{s}(t) = b \cdot q(t) - \left\{ \sum_{i=1}^{n} c_i \cdot \delta_-(t - t_i) \right\} \cdot s(t) \tag{8.34}$$

The magnitude of the ith hit, expressed as a fraction of current $s(t)$, is given by c_i and the times at which the hits occur are given by t_i. The "minus" subscript to the delta function indicates that the ith hit is applied to the value of $s(t)$ just prior to t_i. Although hits may occur during both the acquisition period and during operational use, expenditures are assumed to be zero during operational use (hence $q(t)$ is zero at times greater than t_f), as in previous sections. Then, the objective function is:

$$J = K \cdot \int_{t_f}^{t_L} s(t) \cdot dt - \int_{0}^{t_f} q^2(t) \cdot dt \ = K \cdot s(t_f) \cdot h(t_f) - \int_{0}^{t_f} q^2(t) \cdot dt \tag{8.35}$$

where:

$$h(t_f) = \left[(t_{m(t_f)+1} - t_f) + \sum_{i=m(t_f)+1}^{n} \left((t_{i+1} - t_i) \cdot \prod_{j=m(t_f)+1}^{i} (1 - c_j) \right) \right] \tag{8.36}$$

The parameter $m(t_f)$ is the index of the last hit between time 0 and t_f, that is, the last hit during the acquisition period, and $m(t_f)$ is zero if there are no hits during the acquisition period. The summation in Equation (8.36) is defined to be zero if the starting index exceeds the final index. Furthermore, $t_0 \equiv 0$ and $t_{n+1} \equiv t_L$. Thus, Equation (8.36) accounts for instantaneous changes in the value of $s(t)$ during the operational use period, due to hits specified in Equation (8.34). If no hits occur during the operational use period, then Equation (8.35) reduces to Equation (8.5), the base model objective function.

---

* Wojcik, L.A. and K.C. Hoffman. 2007. ©2007 IEEE. IEEE/SMC International Conference on System of Systems Engineering (SoSE), April 16–18. With permission.

The optimal control solution to Equations (8.34) through (8.36) from the necessary conditions in Table 8.1 is more complex than in the previously described cases, and it was most convenient to generate the closed-form solution for optimum $q(t)$ Equation (8.37), but to find the optimum value of t_f numerically. The numerical procedure was to compute the optimum $q(t)$ across a range of t_f values between time 0 and t_L, and identify the value of t_f that maximizes J. In the case corresponding to results shown later in this section, 60 values of t_f, uniformly spaced between 0 and t_L, were found to be adequate for this purpose. The optimal control $q(t)$ exhibits step changes at hit times during acquisition, and $s(t)$ shows sawtooth-type behavior during acquisition (because $s(t)$ increases linearly between hits and instantaneously changes at hit times) and step changes during operational use (because expenditures are assumed to be zero, but hits continue to affect $s(t)$).

$$q(t) = \begin{cases} \dfrac{K \cdot h(t_f) \cdot b}{2} \cdot \displaystyle\prod_{j=1}^{m(t_f)} (1 - c_j) & \text{for} \ \ t_{j-1} \le t \le t_j \ \ \text{and} \ \ 0 \le j \le m(t_f) \\[2ex] \dfrac{K \cdot h(t_f) \cdot b}{2} & \text{for} \ \ t_{m(t_f)} \le t \le t_f \end{cases}$$

(8.37)

Figure 8.5 shows an example set of hits whose magnitude decreases as a function of time for a system with a total lifetime t_L of 15 years. The time (t_i) of each hit is indicated along the time axis with a nearly vertical spike, and the height of each spike indicates the magnitude, corresponding to the constants c_i in Equation (8.34).

$$c(t) = \prod_{i=1}^{n} c_i \cdot \delta(t - t_i)$$

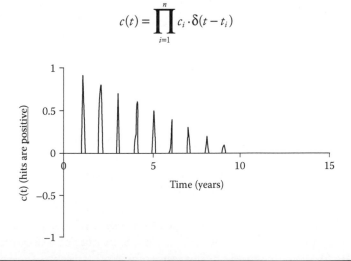

Figure 8.5 A SE scenario involving hits. (From Wojcik, L.A. and K.C. Hoffman. 2007. ©2007 IEEE. IEEE/SMC International Conference on System of Systems Engineering (SoSE), April 16–18. With permission.)

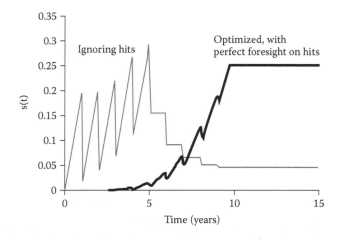

Figure 8.6 SE behavior with life-cycle hits. (From Wojcik, L.A. and K.C. Hoffman. 2007. ©2007 IEEE. IEEE/SMC International Conference on System of Systems Engineering (SoSE), April 16–18. With permission.)

In Figure 8.6, the lighter-shaded curve shows $s(t)$ for the hit scenario indicated in Figure 8.5, with the controls $q(t)$ and t_f optimized by ignoring hits [i.e., using Equations (8.4) and (8.5)]. The bold curve shows $s(t)$, with control $q(t)$ optimized with perfect foresight of the hits shown in Figure 8.5, using Equation (8.37) and numerically optimizing on t_f.

Ignoring (or unaware of upcoming) hits, the SE program experiences wild swings in $s(t)$, followed by a period of operational use in which $s(t)$ suffers further declines. The $s(t)$ profile that optimizes J with perfect foresight of upcoming hits begins much more slowly and gradually rises to a higher level of operational value. The period of relatively slow increase in $s(t)$ corresponds to very limited development until operational requirements are stable, instead of full-scale acquisition in which changing requirements create havoc on the program.

As Figure 8.7 shows, the optimum total expenditures profile (the integral of $q^2(t)$ from time zero to the time on the horizontal axis, indicated by the bold curve) is much more gradual than if hits are not accounted for (the lighter-shaded curve). Taking into account hits in this example, the optimum time to acquire the system is 9.75 years, compared to the 5-year time to acquire ignoring hits, consistent with the one-third rule discussed above.

Thus, we observe that, with foresight of upcoming hits to a SE program, it may be best to begin cautiously, and keep the development relatively limited. Many other hits and helps were run with the model. In some scenarios where hits increase in size over time, it can be best to build a small system rapidly, suggesting a build-a-little, test-a-little approach. When helps, rather than hits, dominate the SE program, corresponding to new operational applications of the system, then aggressive development is called for. In this way, the model can be used to characterize different

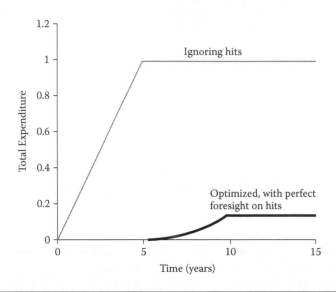

Figure 8.7 Total expenditures as a function of time. (From Wojcik, L.A. and K.C. Hoffman. 2007. ©2007 IEEE. IEEE/SMC International Conference on System of Systems Engineering (SoSE), April 16–18. With permission.)

strategy regimes for SE management and governance. The model presented here assumes full knowledge of upcoming hits and helps, which will not prevail in real SE programs, although in many cases probabilistic estimates of hit magnitudes and times throughout the program life cycle can be generated based on past experience with similar programs. The model can be used to illustrate outcomes across a range of different strategy regimes and to show a perfect-foresight strategy for any set of hits and helps during the system lifetime. A single scenario, including computation of optimum strategy runs in less than one second on a typical, recent laptop computer, so many scenarios can be played out. A possible extension of the model, which has not yet been undertaken, would be to include stochastic simulation to account for uncertainty in number, times, and magnitudes of hits and helps, as well as risk preferences of decision makers.

8.4 Differential Game Models

8.4.1 Linear Differential Game of Self-Interested, Synchronizing Stakeholders

This section describes a model of interacting stakeholders that influence an SE process connected with an enterprise. Such stakeholders may include contractors, government organizations, system user groups, unions, or myriad other individuals or groups pertaining to the enterprise. It is assumed here that stakeholders need to "synchronize" in some fashion for the SE process to run well. For example, in some large military SE programs, sets of contractors need to collaborate, share information, and coordinate their development activities toward a system-level capability. In another current major civil aviation SE program, called NextGen, a large set of stakeholders, including profit-seeking airlines, government agencies, and airport organizations, needs to reach and maintain consensus, and provide political support toward a final system involving many interrelated programs implemented in many geographic areas across the United States (FAA 2010). Although the specific details of these programs vary widely, they all have multiple stakeholders with diverse interests who need to synchronize in some way to facilitate a successful SE result. The modeling presented here seeks to represent the essential features of stakeholder synchronization in SE programs.

In the modeling, the perspective is taken that individual stakeholders are each motivated by their self-interests, which may vary widely across the stakeholder population. Furthermore, it is assumed there is an enterprise agent that can apply incentives and other actions to influence other stakeholders. In the case of a government-led enterprise, this might be a government policy-making organization. For a profit-making firm it could be a high-level executive or organization in charge of the SE effort. For simplicity, the model has each stakeholder making a limited set of decisions as a function of time: at minimum, each stakeholder must decide how much to synchronize with each of the other stakeholders. Each stakeholder can directly benefit from greater synchronization through incentives provided by the enterprise, and perhaps through other mechanisms as well, including the operational benefit to a particular stakeholder of a better final system achieved through greater synchronization. However, synchronization entails cost to stakeholders, related to the effort applied to synchronization activities such as meetings and meeting preparation, and this cost can be substantial, particularly because highly skilled technical people are often needed to perform the activities. Synchronization cost also includes costs associated with possible exposure of internal information to other stakeholders, although these costs may be negative as well for stakeholders who receive internal information about other stakeholders during synchronization activities. The interaction of stakeholder decision making over time, based on benefits and costs seen by stakeholders over the lifetime of the system (including both the SE process and operational use) is modeled here with a simple dynamic game.

A familiar "solution" concept for a dynamic game is the Nash equilibrium (Nash 1950). In a Nash equilibrium, no stakeholder has incentive, based on self-interest, to change its decision unilaterally. In this chapter, we emphasize a weaker version of equilibrium, called here a *local Nash equilibrium*, characterized by each stakeholder having no incentive to make incremental unilateral changes to the amount it synchronizes with other stakeholders. Equilibrium is further specialized to open-loop equilibrium, which specifies the behaviors of the stakeholders over time, rather than a closed-loop feedback specification.

Even more than the single-stakeholder modeling presented before, models of self-interested, synchronizing stakeholders are not expected to be rigorously predictive. Too many factors influence individual stakeholder behavior to expect a model to predict behavior accurately. So, we emphasize the general characteristics of stakeholder behavior that emerge from the modeling. Equilibrium solutions to games involving stakeholders are best understood as potential "attractors" for behavior, rather than strict predictions.

8.4.1.1 Tragedy of the Commons

An obvious incentive scheme is to reward each stakeholder according to the final result of the SE process, achieved across all stakeholder efforts and presumably enhanced by greater synchronization among the stakeholders. A similar case is where the stakeholders benefit directly from the final system and hence may be motivated to synchronize and otherwise contribute to the quality of the final system.

With such an incentive scheme, an individual stakeholder is rewarded for its contribution to the final result of the SE process, but that stakeholder does not receive the rewards that go to the other stakeholders. This is an example of an economic "externality"; that is, each stakeholder does not see the full benefit of its synchronization. Externalities are seen in other domains such as ground transportation, where the typical effect is overuse of a limited resource. Hence, it is sometimes referred to as the "tragedy of the commons" (TOC) (Hardin 1968). In the case of SE stakeholders, the effect is *under-synchronization* across the set of stakeholders, despite the fact that each stakeholder decides in its own self-interest. The under-synchronization occurs because each stakeholder does not receive the benefit that goes to other stakeholders from its own synchronization; hence, all stakeholders tend to do better if all synchronized more than they do at equilibrium.

There are well-known mitigations of TOC to prevent overuse of a limited resource, including some that efficiently internalize the costs of economic externalities (Ellis and van den Nouweland 2000). TOC mitigations include establishing central control mechanisms (e.g., centrally planned transportation systems [Cantarella and Sforza 1991]) or penalties to remove economic externalities (e.g., congestion pricing for heavily used highways), and providing mechanisms for resource users to reach binding agreements not to overuse the limited resource. The analogies for stakeholder synchronization are threefold.

First, stakeholders can be constrained to synchronize with each other by the enterprise. For example, in the case where the stakeholders consist of a set of contractors, contracts can be written to require synchronization. This approach suffers from the great disadvantage of limited adaptability: as the SE program proceeds and conditions change, requirements for synchronization, as originally conceived and written into contracts, may cease to be helpful. Furthermore, it is not feasible at all to bind many kinds of stakeholders through contractual or other legal arrangements.

Second, stakeholders can be provided with additional incentives to synchronize further. These can be financial or nonfinancial incentives provided by the enterprise. However, at some point the cost of such incentives will exceed the benefits of providing them. Modeling of stakeholder incentive provision leads to another set of behavioral patterns discussed later in this chapter.

Third, an environment can be fostered that facilitates a "cooperative game" among stakeholders, with effectively "binding agreements" between stakeholders to help ensure they will not defect from a position of higher synchronization than is in their immediate self-interest. This could be accomplished through an "enterprise culture" that rewards good synchronization behavior and punishes bad behavior. These rewards and punishments need not be financial, and can be introduced at any level of an organization, including individuals in the organization. For example, special recognition can be given to individuals who are especially active in coordinating across organizations to achieve synchronization. And, if opportunities to synchronize occur repeatedly throughout the SE program, or across multiple SE programs, then good synchronization can be promoted through a "tit for tat" threat across time (Axelrod 1984).

8.4.1.2 Multiple Equilibria

Another general characteristic of many types of games is the existence of multiple equilibria. That is, the solution to the game is not unique. Using the specific instantiation of a stakeholder synchronization game described in the next section, such multiple equilibria have been observed. Interpreting each equilibrium as an "attractor" for collective behavior, multiple equilibria means that collective behavior may be attracted in multiple directions, which may depend upon seemingly insignificant or capricious factors that might direct collective behavior in one direction as opposed to another. In this sense, there are fundamental limits to predictability of stakeholder systems, even ones for which extensive and detailed information about motivations and costs and benefits is available, and in which stakeholders behave "rationally".

8.4.2 Enterprise Incentives in a Stackelberg Game of Stakeholders

In the previous section, general characteristics of equilibrium solutions to a game of synchronizing stakeholders were discussed. To address incentives for stakeholder synchronization, the synchronization game is expanded as depicted to include a set of interacting stakeholders as well as an enterprise policy agency to provide incentives. In this chapter, provision of enterprise incentives is modeled with a leader–follower, or Stackelberg game (Fudenberg and Tirole 1991), in which the leader is the enterprise agency providing incentives and the followers are the set of stakeholders who each decide on a level of synchronization in their own self-interest. The approach to solve such a game is to work backwards: first, the stakeholders are assumed to be in equilibrium, then the incentives from the enterprise agency are optimized so as to maximize benefit minus cost to the enterprise.

The model proposes a value to stakeholder i from a set of decisions across all stakeholders and time as:

$$v_i\left(u_1(t), \ldots, u_n(t), r_i(t)\right) = h_i \cdot v_s\left(u_1(t), \ldots, u_n(t), r_i(t)\right)$$

$$-d_i \cdot \int_0^{t_f} u_i(t) \cdot dt - f_i \cdot \int_p^{t_f} r_i(t) \cdot dt \tag{8.38}$$

In the above equation, the value of the system to the enterprise is:

$$v_s\left(u_1(t), \ldots, u_n(t), r_1(t), \ldots, r_n(t)\right) =$$

$$\int_0^{t_f} \left\{ \sum_{j=1}^{n} \left[a_j \cdot g\left(u_j(t)\right) \right] + b \cdot g\left(\sum_{j=1}^{n} r_j(t) \middle/ n \right) \cdot \sum_{j=1}^{n} g(u_j) \right\} \cdot dt \tag{8.39}$$

And,

$$g(x) = (1 - \exp(-x))/(1 - \exp(-1)) \tag{8.40}$$

In Equations (8.38) and (8.39), $u_i(t)$ represents the internal effort expended by stakeholder i toward the enterprise capability, and $v_i(t)$ is the synchronization effort towards the capability. We assume $u_i(t)$ and $v_i(t)$ have maximum values of 1. The variables, d_i, f_i, a_i, and b are constant in time. The constant n is the number of stakeholders. The first term on the right-hand side (RHS) of Equation (8.38) is the incentive provided by the enterprise, and the i that maximizes the value of the system to the enterprise. The second and third terms represent the cost of internal and synchronization efforts, integrated over time.

In Equation (8.39), the first term in the integrand on the RHS is the contribution to total enterprise value from the internal efforts of the stakeholders, assumed here simply to add. Following a model by Huberman and Glance (1998), we assume the value from synchronization to be proportional to the product of synchronization terms among the stakeholders. Furthermore, it is assumed that this term is proportional to the product of individual efforts. In other modeling work, we have sometimes assumed a sum of products across partial sets of stakeholders, which would correspond to system value accruing from synchronization across partial sets of stakeholders, but in the presentation here we assume value accrues from synchronization across the entire set of stakeholders.

It was found most convenient to generate solutions to Equations (8.38) through (8.40) numerically using the method of fictitious play, although we apply necessary conditions (Bryson and Ho 1975, p. 277) to demonstrate that at equilibrium, both internal effort and synchronization effort of each stakeholder is constant in time. This follows from conditions on the Hamiltonian,

$$\partial H_i / \partial u_i = 0 \tag{8.41}$$

$$H_i = \lambda_i(t) \cdot u_i(t) + \mu_i(t) \cdot r_i(t) + L_i(u_i(t), r_i(t), t) \tag{8.42}$$

$$\partial H_i / \partial r_i = 0 \tag{8.43}$$

$$\dot{\lambda}_i = -\partial H_i / \partial s_i \tag{8.44}$$

$$\dot{\mu}_i = -\partial H_i / \partial w_i \tag{8.45}$$

Here,

$$\dot{s}_i(t) = u_i(t) \tag{8.46}$$

$$\dot{w}_i(t) = r_i(t) \tag{8.47}$$

In Equations (8.41) through (8.45), $\lambda_i(t)$ and $\mu_i(t)$ are Lagrange multiplier functions across stakeholders and the Lagrangian is:

$$L_i(u_i(t), r_i(t), t) = -d_i \cdot u_i(t) - f_i \cdot r_i(t) \tag{8.48}$$

Note that the Lagrangian has no explicit time dependence and that:

$$\dot{\lambda}_i = \dot{\mu}_i(t) = 0 \tag{8.49}$$

This suffices to show that $u_i(t)$ and $v_i(t)$ are independent of time.

The method, then, for producing numerical results is to fix a set of incentives across the stakeholders, solve for equilibrium among the stakeholders using the method of fictitious play, and then vary the incentives until the maximum system value minus incentive cost is achieved:

$$V = v_s \left(u_1(t), ..., u_n(t), r_1(t), ..., r_n(t) \right) - \sum_{i=1}^{n} h_i \qquad (8.50)$$

8.4.2.1 Hardness Catastrophe in Incentives

The results show a consistent effect termed here a "hardness catastrophe" and illustrated in Figure 8.8. Here, three stakeholders are modeled, and the optimum stakeholder 3 with difficulty to the other stakeholders held constant. As internal difficulty increases, the optimum incentive to stakeholder 3 also increases, until a threshold is reached and the optimum incentive drops to zero. This happens because, as difficulty of the development effort increases, the enterprise spends more on incentives, until the point is reached where the amount spent on incentives reaches the enterprise value of the system and it is no longer worthwhile to provide incentive to stakeholder 3.

This could be a dynamic effect in complex enterprise SE efforts, in which the difficulty of the SE effort is not apparent at the beginning of the program, but gradually becomes more apparent. Wishing for ultimate success, the enterprise provides additional incentives to a contractor that finds more and more difficulty, until the catastrophe threshold is reached, and the contractor is dropped from the program.

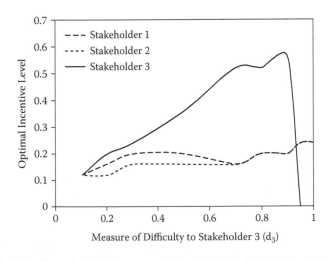

Figure 8.8   Variation in internal difficulty to a stakeholder.

8.4.2.2 Robustness and Program Characteristics

Next we consider the problem of program robustness, or the ability of a program to survive in the face of unexpected changes. Following the results already presented, we focus on robustness with respect to uncertainty about internal difficulty to a stakeholder. One metric for this kind of robustness is the level of difficulty at which the hardness catastrophe occurs for a specific stakeholder, as discussed above. A high level of robustness means the enterprise can adjust incentives to maintain the stakeholder's development activities optimally, even if the internal difficulty to that stakeholder turns out to be high.

An important question for SE programs is: what program characteristics give rise to a high robustness level? Here, we vary the extent to which synchronization among the stakeholders is important to the overall effort, relative to individual effort, keeping the maximum operational value of the program constant. This is done by choosing different values of the constant b and renormalizing the system value as:

$$v_s(u_1(t), ..., u_n(t), r_1(t), ..., r_n(t)) =$$

$$\left(\frac{v_{s,max}}{\sum_{j=1}^{n} a_j + n \cdot b} \right) \cdot \int_0^{t_f} \left\{ \sum_{j=1}^{n} [a_j \cdot g(u_j(t))] + b \cdot g\left(\sum_{j=1}^{n} r_j(t)/n \right) \cdot \prod_{j=1}^{n} g(u_j) \right\} \cdot dt \qquad (8.51)$$

A measure of importance of synchronization can be expressed by the parameter,

$$I = \left(\frac{n \cdot b}{\sum_{j=1}^{n} a_j + n \cdot b} \right) \qquad (8.52)$$

Numerical simulations indicate in Figure 8.9 that there is a middle range of synchronization importance at which robustness is maximized. As described before, the scenarios analyzed here involve three stakeholders, and the internal difficulty to one of the stakeholders is varied. An interpretation of these results is that if the synchronization importance is too high, then the difficulties experienced by stakeholder 3 bring down the other stakeholders, and hence the entire system, relatively soon. At the other extreme, if the stakeholders work essentially independently, then stakeholder 3 does not derive any benefit from the other stakeholders' efforts which might counterbalance its own difficulties. Hence, there is a middle ground where robustness is maximized.

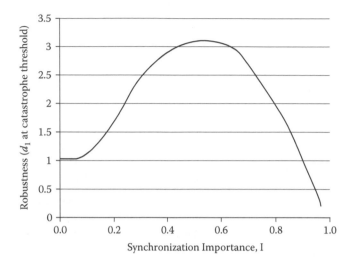

Figure 8.9 Dependence of robustness on synchronization importance.

8.4.3 *Enterprise Coevolution with the External Environment*[*]

This section describes a differential game model to explore the dynamics of a range of enterprise types. Here, the enterprise might correspond to a government agency or set of agencies with a mission, and its environment is the external world that interacts with the enterprise. For example, the enterprise could be the FAA and its environment could be the users of the aviation system, including airlines, general aviation, and military users. Or, the enterprise could be the Internal Revenue Service (IRS) and its environment could be U.S. taxpayers. The enterprise is motivated to create capabilities that meet the demand placed on it by the environment. The environment has its own motivations, which may or may not align with the motives of the enterprise being examined. For simplicity, it is assumed that the respective motivations of the enterprise and its environment can be represented as costs per unit time given by the following pair of equations:

$$L_k = A \cdot (d(t) - k(t))^2 + C \cdot d(t) + B \cdot \dot{k}(t)^2 \tag{8.53}$$

$$L_d = D \cdot (d(t) - k(t))^2 - E \cdot d(t) + F \cdot \dot{d}(t)^2 \tag{8.54}$$

---

[*] Adapted from Wojcik, L.A. and K.C. Hoffman, IEEE/SMC International Conference on System of Systems Engineering (SoSE), (2006), with significant changes. © 2006 IEEE. With permission.

In Equations (8.53) and (8.54), A, C, B, D, E, and F are assumed to be constants. The functions $d(t)$ and $k(t)$ can be assumed to be always nonnegative, representing, respectively, "demand" placed on the enterprise by the environment and "capacity" to meet the demand produced by the enterprise. Here, "demand" and "capacity" are understood very generally, so they apply across many kinds of enterprises and corresponding environments. In this simple model, we assume the first derivatives of $d(t)$ and $k(t)$ are always nonnegative. The objective of the enterprise player is to minimize its cost L_k integrated over a given time interval from T_0 to T_F. Similarly, the objective of the environment player is to minimize its cost L_d integrated over the same time interval. It is possible to extend this formulation of the game so that the enterprise and the environment have different lookahead times, but that is not presented here.

In Equations (8.53) and (8.54), the first term on the right-hand side represents the interaction between the enterprise and the environment. It is assumed that the constant A is positive, so that the enterprise is motivated to produce operational capabilities $k(t)$ that tend to "match" the demand $d(t)$ imposed by the external world. The constant D, however, may be positive, zero, or negative based on the type of enterprise and environment.

The second term on the right-hand side of Equation (8.53) represents the burden placed on the enterprise by the demand $d(t)$ produced by the external environment, even if the capacity of the enterprise is matched to the demand. Meeting this burden may entail providing services and capabilities, in a quantity and quality consistent with regulatory and financial constraints placed on the enterprise. Thus, the constant C ordinarily would be nonnegative. Similarly, the second term on the right-hand side of Equation (8.54) represents the benefit to the environment from the demand it imposes on the enterprise. Ordinarily, E is positive, reflecting that the environment derives benefit from its operations, so E is positive.

Finally, the third terms in both Equations (8.53) and (8.54) represent the difficulty of generating capability in the case of the enterprise and demand in the case of the environment. The more difficult (and costly) it is to generate capability per unit time the larger is the constant B, and similarly for demand and the constant F.

The cost Equations (8.53) and (8.54) can be generalized to make the terms they contain completely symmetric (allowing for unequal numerical values of the corresponding constants), which may be appropriate for modeling a wider range of enterprise types; for example, the U.S. DoD acting against various environments, ranging from the past cold-war era Soviet Union through the current asymmetric warfare threat, but this generalization is beyond the scope of this chapter.

We solve the game for game-theoretic equilibriums in open-loop form with initial conditions specified as $k(0) = k_0$ and $d(0) = d_0$. An open-loop Nash solution specifies $k(t)$ and $d(t)$ as functions of time such that neither player has unilateral incentive to change. The intuitively suitable perspective can be taken that the cost functions are the controls for Equations (8.53) and (8.54), but it is mathematically equivalent to take the first derivative of the state variable in each equation as the control variable. Then, the equations of state are trivial:

$$\dot{k}(t) = \dot{k}(t) \tag{8.55}$$

$$\dot{d}(t) = \dot{d}(t) \tag{8.56}$$

Applying the necessary conditions for open-loop equilibrium solutions (Bryson and Ho 1975) where constraints are slack, there emerge three types of solutions, depending on the value of the constant parameter,

$$G = \frac{D}{F} + \frac{A}{B} \tag{8.57}$$

When G is greater than zero, potential solutions $k(t)$ and $d(t)$ are linear combinations of positive and negative exponential functions of time with time constant equal to the square root of G, and second-degree polynomials in time. When G equals zero, potential solutions are fourth-degree polynomials in time. When G is less than zero, potential solutions are linear combinations of sinusoidal functions and second-degree polynomials in time, with sinusoidal angular frequency given by:

$$\omega = \sqrt{-G} \tag{8.58}$$

Figure 8.10 shows an example solution with $A = B = C = D = E = F = 1$ with $d_0 = 5$ and $k_0 = 3$, and with $T_0 = 0$ and $T_F = 10$. In this example, $G > 0$. The enterprise and the environment favor matching enterprise capacity to environmental demand, and they tend to coadapt in ways that reduce the difference between capacity and demand. The FAA enterprise and its environment of airspace system users probably corresponds to such a "cooperatively adaptive" enterprise/environment system. When $D = 0$, demand $d(t)$ is insensitive to enterprise capacity, so its time evolution is independent of $k(t)$. However, the enterprise adapts to changes in demand produced by the environment. The IRS enterprise and its environment probably can be approximated by this "insensitive demand" regime. Coevolution of demand and capacity are qualitatively similar to that of cooperatively adaptive enterprise/environment systems. A qualitatively different behavioral regime is where enterprise and environment are "oppositional," with $D < 0$, which is expected for the DoD/DHS enterprise and its environment of asymmetric threats. In this regime, the enterprise favors matching its capability with the demand of the external environment, but the environment favors maximizing the difference between enterprise capacity and environmental demand. In this regime, enterprise capabilities and

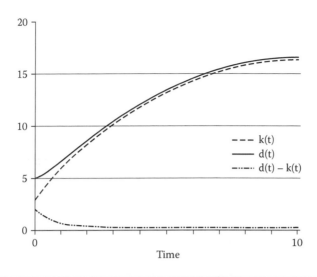

Figure 8.10 **Modeled coevolution of cooperatively adaptive enterprise and environment. (From Wojcik, L.A. and K.C. Hoffman. 2006. ©2006 IEEE. IEEE/SMC International Conference on System of Systems Engineering (SoSE). April 24–26. With permission.)**

environmental demand may both increase very rapidly with time, with their difference also increasing with time. For parameter values of $A = B = C = E = F = 1$ with $d_0 = 5$ and $k_0 = 3$, and with $T_0 = 0$ and $T_F = 10$, Wojcik and Hoffman (2006) report a critical value of $D_c = -1.02467$, at which the rates of increase approach infinity in this simple model. The simple model does not include many real-world constraints, which of course do not permit infinite rates.

8.5 Conclusion

Instantiations of the framework presented here for modeling the SE process show behavior patterns that can be compared to real SE programs. Quantitative validation of these models is likely to be limited, but there is nevertheless potential for the models to be used for education, information sharing, SE program diagnostics, and early-stage screening of potential SE management and governance approaches.

Specific modeling results of interest include the "one-third" rule for development time, relative to total system lifetime, of systems with moderate internal complexity, and deviations from the one-third rule for systems with other levels of internal complexity, as well as deviations due to such effects as system aging and external complexity. Systems developed in stages also have characteristic optimal development times relative to total system lifetime, and the "HOT" model originally developed (Wojcik 2004) to model the systems engineering process is shown to be a limiting case of the more general framework, with infinite incremental stages of development. Modeling of self-interested enterprise stakeholders in system development shows the familiar phenomena of the "tragedy of the commons" and multiple equilibria. In addition, the modeling produces a "hardness catastrophe" in incentives as a stakeholder's internal difficulty is increased. And, simulation results point to a "sweet spot" in internal synchronization importance in terms of maximizing the robustness of the multistakeholder enterprise SE process. Finally, it is demonstrated that different types of enterprise–environment relationships, ranging from cooperatively adaptive to oppositional, can produce qualitatively different behaviors. In the simple model presented here, oppositional relationships can produce singular behavior with respect to a key parameter expressing the degree of oppositional motivation of the environment.

The modeling presented here is intended to inspire more investigation. Much more research is needed to link the models to real SE programs and real systems, to apply the models to improve new and ongoing SE programs, and to demonstrate the practical utility of the models. On a broader level, perhaps the high-level optimal control and differential game modeling of the SE process presented here could lead to deeper understanding of complex SE and human/technological systems in general.

References

Alderson, D.L. and Doyle, J.C. 2010. Contrasting views of complexity and their implications for network-centric infrastructures. *IEEE Transactions on Systems, Man and Cybernetics—Part A: Systems and Humans,* 40(4, July): 839–852.

Alderson, D.L, Li, L., Willinger, W., and Doyle, J. 2004. The role of design in the Internet and other complex engineering systems. Presented at the New England Complex Systems Institute (NECSI) International Conference on Complex Systems (ICCS), Boston.

Arthur, W.B. 2009. *The Nature of Technology: What It Is and How it Evolves.* New York: Free Press.

Axelrod, R. 1984. *The Evolution of Cooperation.* New York: Basic.

Bar-Yam, Y. 2003. When systems engineering fails—Toward complex systems engineering. International Conference on Systems, Man & Cybernetics, Vol. 2: Piscataway, NJ: IEEE Press, pp. 2021–2028.

Boppana, K., Wang, Z., Wheeler, P., and Zborovskiy, M. 2005. AAS: Comparison of HOT model and system dynamics model. Briefing presented at *Massachusetts Institute of Technology* (December).

Boppana, K., Wojcik, L. et al. 2006. Can models capture the complexity of the systems engineering process? Presented at the New England Complex Systems Institute (NECSI) International Conference on Complex Systems (ICCS2006), Boston, 25–30 June.

Bryson, A.E. and Ho, Y.-C. 1975. *Applied Optimal Control.* Washington: Hemisphere.

Cantarella, G.E. and Sforza, A. 1991. Traffic assignment. In M. Papageorgiou (Ed.) *Concise Encyclopedia of Traffic and Transportation Systems.* Oxford: Permagon Press, 513–520.

Carlock, P.G., Decker, S.C., and Fenton, R.E. 1999. Agency-level systems engineering for "systems of systems". *Systems and Information Technology Review Journal*, Spring/Summer: 99–110.

Carlson, J.M. and Doyle, J. 1999. Highly optimized tolerance: A mechanism for power laws in designed systems. *Physical Review E*, 60(2, August): 1412–1427.

Department of Transportation, Office of the Inspector General. 1998. Audit Report, Federal Aviation Administration, Advance Automation System. AV-1998-113 (April 15). http://www.oig.dot.gov/sites/dot/files/pdfdocs/av1998113.pdf

Ellis, C.J. and van den Nouweland, A. 2000. A mechanism for inducing cooperation in non cooperative environments: Theory and applications. *Social Science Research Network* (February). http://papers.ssrn.com/sol3/papers.cfm?abstract_id=436522

FAA (Federal Aviation Administration). 2010. FAA's NextGen Implementation Plan (March). Washington, DC: Federal Aviation Administration.

Fudenberg, D. and Tirole, J. 1991. *Game Theory*. Cambridge, MA: MIT Press.

Hardin, G. 1968. The tragedy of the commons. *Science*, 162: 1243–1248.

Hoffman, K.C. et al. 2007. Descriptive enterprise dynamics—A multi-disciplinary unifying framework. Presented at the Fifth Annual Conference on Systems Engineering Research (CSER), International Council on Systems Engineering (INCOSE), Hoboken, NJ.

Huberman, B.A. and Glance, N.S. 1998. Fluctuating efforts and sustainable cooperation. In M.J. Prietula et al. (Eds.), *Simulating Organizations: Computational Models of Institutions and Groups*. Menlo Park, CA: American Association for Artificial Intelligence.

Nash, J.F. 1950. Equilibrium points in n-person games. *Proceedings of the National Academy of Sciences*, 36 (1): 48–49.

Popper, S.L., Lempert, R.J., and Bankes, S.C. 2005. Shaping the future. *Scientific American* (April): 66–71.

Stubna, M.D. and Fowler. J. 2003. An application of the highly optimized tolerance model to electrical blackouts. *International Journal of Bifurcation and Chaos*, 13(1): 237–242.

U.S. Congress, Office of Technology Assessment. 1982. Review of the FAA 1982 National Airspace System Plan, Library of Congress Catalog Number 82-600595, Washington, DC: U.S. Government Printing Office.

Wojcik., L.A. 2004. A highly-optimized tolerance (HOT) model of the large-scale systems engineering process. In *Student Papers: Complex Systems Summer School*, June 6–July 2, Santa Fe, NM: Santa Fe Institute.

Wojcik, L.A. and Hoffman, K.C. 2006. Systems of systems engineering in the enterprise context: A unifying framework for dynamics. Presented at the 2006 IEEE/SMC International Conference on System of Systems Engineering (SoSE), April 24–26 (Digital Object Identifier 10.1109/SYSOSE.2006.1652268).

Wojcik, L.A. and Hoffman, K.C. 2007. Emergent enterprise dynamics in optimal control models of the system of systems engineering process. Presented at the 2007 IEEE/SMC International Conference on System of Systems Engineering (SoSE,), April 16–18.

Chapter 9

Hybrid Systems Dynamic, Petri Net, and Agent-Based Modeling of the Air and Space Operations Center

Jennifer Mathieu, John James, Paula Mahoney, Lindsley G. Boiney, Richard Hubbard, and Brian E. White[*]

Contents

---

[*] Adapted from Mathieu, J. et al. 2007a. Hybrid system dynamic, Petri net, and agent-based modeling of the Air and Space Operations Center. In INCOSE Symposium, Systems Engineering: Key to Intelligent Enterprises, June 24–28, San Diego.

215

9.1 Introduction

In an earlier paper (Mathieu et al. 2007b), an existing Air and Space Operations Center (AOC) process model (i.e., Petri net) and new global and mission models for the environment in which the AOC operates (i.e., system dynamics) were linked (federated). The focus of this chapter is the development of an operator–environment model (i.e., agent-based model). An existing systems framework for attention allocation of operators within the AOC has been implemented that supports multiple modeling paradigms. The results for linking the Petri net and system dynamics models are summarized, and new results for the agent-based model are presented based on a pilot-down scenario. It has been observed that many AOC operators can become distracted by a pilot-down critical event, even if the operator is not able to assist in the rescue directly. Furthermore, this distraction has been hypothesized to have a detrimental effect on the activities the uninvolved operators are currently handling.

9.2 Background

A set of interrelated activities (or regimen) for complex-system engineering has been suggested (Kuras and White 2005; White 2005a; White 2005b; Kuras and White 2006), including: (1) analyze and shape the environment; (2) tailor developmental methods to specific regimes and scales; (3) identify or define targeted outcome spaces; (4) establish rewards (and penalties); (5) judge actual results and allocate rewards; (6) formulate and apply developmental stimulants; (7) characterize continuously; and (8) formulate and enforce fitness regulations (policing).

These activities were hypothesized to focus and accelerate the natural evolution for the potential benefit of the system. The goal of this effort is to develop a multiscale hybrid model using real data and subject matter experts (e.g., Figure 9.1). The main research question is how to link (federate) the models from the various scales together. Once a multiscale model exists, it will be possible to apply more fully the activities listed above. As part of this study, analyzing and shaping the environment, tailoring development methods, judging results, characterizing continuously, and enforcing fitness regulations were all part of the modeling process. The remaining methods can be applied after the multiscale model is validated.

The earlier paper (Mathieu et al. 2007b), used an existing AOC process model, and developed global and mission models for the environment in which the AOC operates (Figure 9.1). The models were developed in separate software environments, MSim and Vensim®, respectively, and the main challenge was linking the models. The Petri net, AOC process model used discrete time whereas the system dynamics, global, and mission models used continuous time. In addition, the Petri net model was validated using data over the course of one day (1990 Gulf War), and

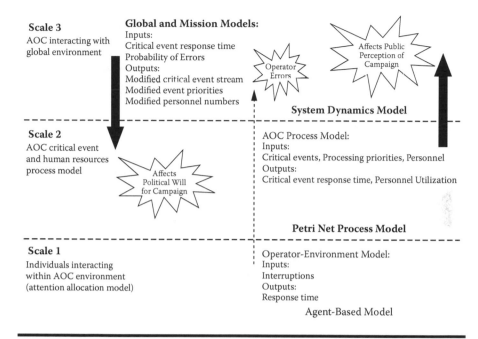

Figure 9.1 Multiscale hybrid model of the AOC with possible linking method.

the system dynamics model was designed to run over a period of 60 days. Finally, model elements were needed to link the Petri net and system dynamics models, and these were created using Vensim because of its rapid model development capabilities. However, this required the re-creation of partial abstractions of the AOC model in Vensim. The results of this analysis are presented briefly below.

The operator–environment model is the focus of this chapter. AnyLogic® was selected as the software environment in which to build the agent-based model; this environment can support multiple modeling paradigms in the same tool, thus addressing the difficulty in linking the models. AnyLogic has the following features that facilitate multiscale hybrid modeling:

■ Supports system dynamics, discrete-event, agent-based, and dynamic systems modeling while using the same timing engine
■ Maintains discrete or continuous space
■ Easy to adjust the level of abstraction
■ Possible to choose the best approach for the problem
■ Easy to switch from one approach to another
■ Easy to mix approaches to develop better models
■ Based on Java, object-oriented
■ Viewer/debugger, stand-alone Java application, Java applet, XML format Java remote method invocation (RMI), high-level architecture (HLA)

9.3 Operator-Environment Model (Agent-Based)

Boiney (2007) describes a systems framework for attention allocation of operators within the AOC, as shown in Figure 9.2. This framework was developed using direct field observations of operators in command and control environments (e.g., Joint Expeditionary Forces Experiment, JEFX '06). The focus of the agent-based modeling effort is the Attention Allocation System shown in the center of the figure. In this first iteration of the model, agents are assumed to focus on only one activity at a time. Therefore, the Primary Attention and Secondary Attention activity boxes shown in Figure 9.2 are modeled as a single operator focus. The various activities that the operator must handle have priority and difficulty (e.g., related to stress) levels and can spawn additional self-activities (e.g., related to memory) such as completing a checklist.

This interpretation can be visualized as an activity queue (Figure 9.3), with the top activity being the current focus. As the agent is interrupted, the order of the activities in the queue can change. If operators are interrupted frequently, the delay in reorienting to new activities can be increased (e.g., due to stress or memory limitation). Agents can be interrupted by various communication modalities including chat, e-mail, telephone, and in-person, with a notional representation of the disruptiveness of each modality shown in Figure 9.3. Subject matter experts have provided the following guidance on how much each modality is used (in terms of an average utilization rate) in a typical command and control environment: chat: 75% (text chat: 70% and audio chat: 5%); e-mail: 10%; telephone: 5%; and in-person: 10%. The agent behavior will also depend on the agent's past experience with the source of the interruption. Certain behavior is expected, depending on whether the source is trustworthy (McCarter and White 2007), for example.

The information itself will also affect agent behavior: it may be specifically related to the activity in which the operator is working, change the operator's priorities, or be a distraction (e.g., media report or pilot-down). Therefore, each interruption is associated with: (1) the agent's past experience with the source of the information; (2) the information type; (3) a communication modality; and (4) whether an attention allocation attractor is present. Attention attractors are defined as elements of higher inherent interest, including being drawn to human faces and voices (especially those indicating emotion), changes in the environment (if not too gradual), informal conversation (versus formal language), bright colors, signs of imminent danger, and anything novel or unexpected or unresolved (Boiney 2007). For example, operators sometimes scan 10 chat rooms when only four are actively used. In this instance, one question is whether it is possible to develop a smaller interface (e.g., summary display) that still provides value to the operator. Such a display can also be designed to provide greater value to the operator through the use of social network software and rule creation (Boiney 2007). The degree of disruptiveness of the attention attractor can span a wide range as illustrated in Figure 9.4.

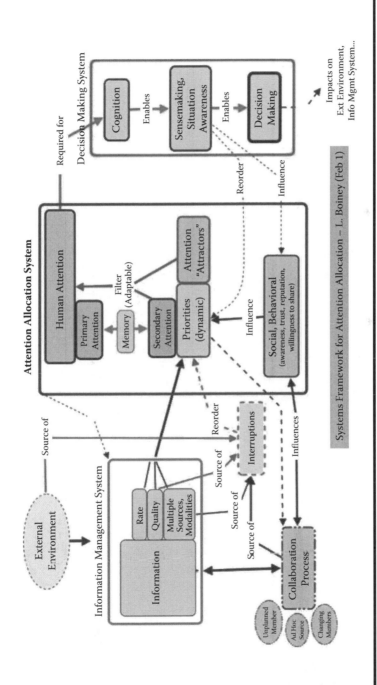

Figure 9.2 Systems framework for attention allocation. (From L.G. Boiney, 12th International Command and Control Research and Technology Symposium (CCRTS) 2007. With permission.)

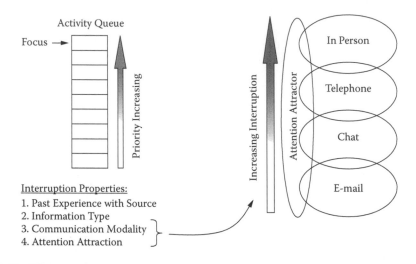

Figure 9.3   Operator activity queue and communication modalities.

The scenario selected for the operator-environment model was the time-sensitive targeting (TST) process, which is modeled as 49 operators working in the processes outlined in Figure 9.4. These operators work in specific groups denoted by the larger boxes that encompass one or more of the individual task boxes. In the current version of the model, one group of 8 operators interacts to complete a task (rectangles in Figure 9.4), based on a random number of activities or subtasks. Each agent is randomly assigned a number of static properties, including a level of expertise, experience, and ease of interaction.

Based on the initial properties, the agent develops a social network as the simulation is run. As shown in Figure 9.5, an agent "grabs" a critical event. The agent checks to see if the event, to be handled properly, needs any more information. If not, the agent is done, and the processing proceeds depending on a random setting of the modality of the next interaction. If the modality is "talking" a check is made to see if the given agent has any trusted associates from the same cell. If so, the agent picks one and talks for a random amount of time. If the modality is "telephone" no associate is picked but a "talk-time" is calculated (assumed outside of cell). In either case, after the talk-time passes the agent checks if success in getting information was accomplished. If the agent is successful (a probabilistic outcome), the amount of information needed to handle an event is decreased, and the agent adds the associate agent to the trusted list (if not already there), and goes back to see if the event needs any more information. If the agent was unsuccessful, a check is made to see whether more information is needed. E-mail and chat modalities are implemented in a simplified fashion.

Agent measures of performance (MOP) or indicators include: transition time between activities or subtasks, errors, missed information, response time, and repeated requests. Potential experiments within the agent-based modeling environment include:

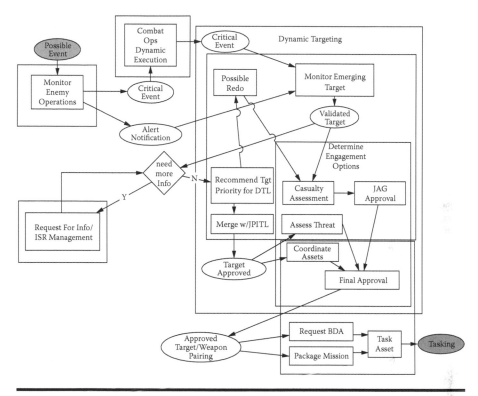

Figure 9.4 Air and space operations center (AOC) process model. The rectangles indicate tasks and arrows indicate the flow of critical events: dynamic target list (DTL), joint integrated prioritized target list (JIPTL), judge advocate general (JAG), and battle damage assessment (BDA).

- Add AOC environment distractions (e.g., media reports or pilot-down).
- Use of attention attractors (e.g., summary display).
- Incentives that motivate operators to collaborate between cells.
- Remove well-contacted operators (i.e., how important are personal relationships?).
- Create environment that has more remote cell members (e.g., distributed AOC).
- Create procedures that increase the chance of bottlenecks (e.g., require all information to pass through one agent).

A pilot-down scenario was developed based on the observation that many AOC operators can become distracted by a pilot-down even if the operator is not able to assist in the rescue directly. Furthermore, this distraction has been hypothesized to have a detrimental effect on the activities being handled by the uninvolved operator. Figure 9.6 shows the agent environment (Java main class) and the pilot-down event timer as well as the timer for other TST events. Circles indicate global variables and squares indicate Java classes (e.g., state charts for operators and chat

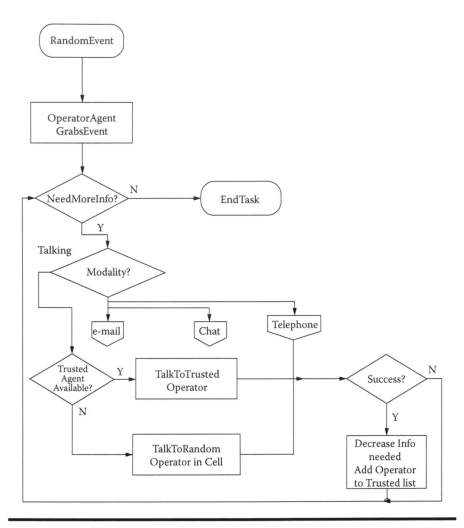

Figure 9.5 Modeling operator activities or subtasks in the agent-based model.

are implemented). Figure 9.6 depicts the agent environment, Figure 9.7 shows the operator class in detail, and Figure 9.8 provides state charts for operator activity and availability.

9.4 AOC Process Model (Petri Net)

The existing AOC process model describes situation awareness and assessment (SA&A) and TST operations, including human resources (Mathieu et al. 2007b). The strike and support missions included six different types of critical events: (1) theater ballistic missile (TBM) launch; (2) TBM detection; (3) combat search

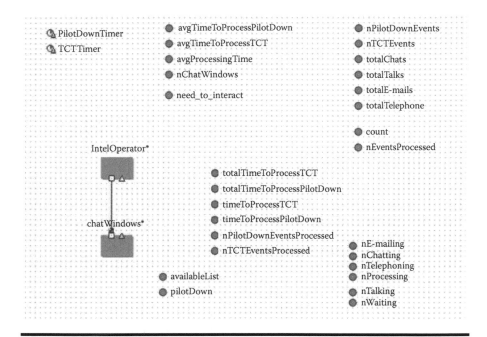

Figure 9.6   The agent environment or Java main class.

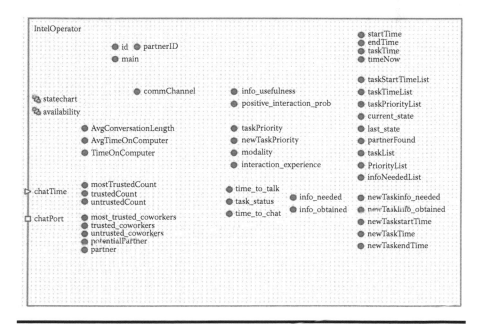

Figure 9.7   The operator class.

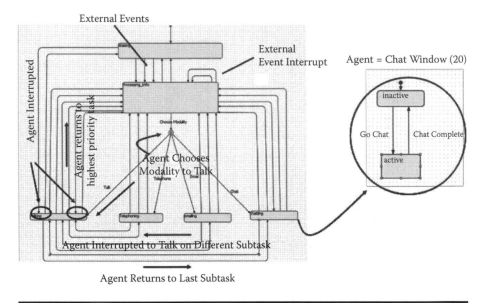

Figure 9.8 **The operator state chart for activities (left) and chat window state chart (right).**

and rescue (CSAR, a pilot-down situation); (4) surface-to-air missile (SAM) radar emissions; (5) choke point (enemy assets constrained by terrain); and (6) air tasking order (ATO) retasking.

All events were given the same priority with the exception of a CSAR event, which was given a higher priority. The effect of this was that the higher priority event pre-empted work on any other event type being processed in the same resource. All events trigger TST responses by the dynamic targeting cell (DTC) of the AOC in addition to their responsibilities of monitoring the progress of the ATO. The focus of this study is on DTC operations during the execution of the campaign as targets of opportunity become available. The process of time-sensitive targeting is outlined in Figure 9.9.

The AOC process model built using MSim (Mathieu et al. 2007b) explores and optimizes the operational processes of the DTC; Figure 9.9 shows a portion of the model. The Petri net model is used to find efficient results for various indicators of performance regarding TST performed in the DTC, including: (1) time from target appearance to target prosecution, critical event response time; (2) workload in DTC, resource utilization; and (3) number of operators in DTC. However, making the DTC as efficient as possible without considering global environment factors may lead to a problem where the "local" optimum produces a result at the system scale that is below the "global" optimum.

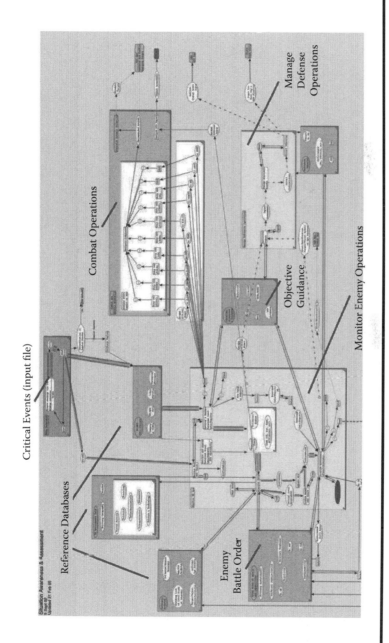

Figure 9.9 **Portion of AOC Petri net model: monitor operations and combat operations, global and mission models (system dynamics).**

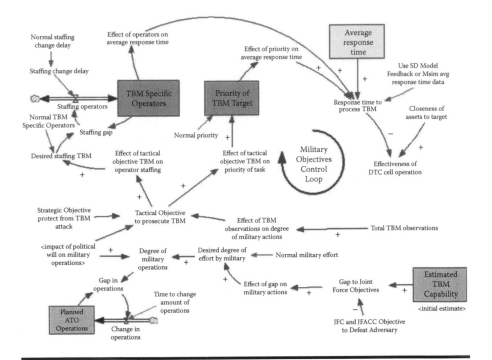

Figure 9.10 System dynamics model of DTC operations for TST of adversary TBMs.

The system dynamics modeling process was used in two distinct manners to support modeling of DTC operations: informing the AOC Petri net model of global-scale dynamics and linking the Petri net and system dynamics models to achieve strategic mission and global-scale simulation. The system dynamics model is composed of four sections (Mathieu et al. 2007b). The input values critical event response time (Figure 9.10) and personnel utilization from the AOC process model are introduced, and behavior derived from the input values is modeled (Figure 9.11).

The system dynamics model can be simulated either by itself, or in conjunction with the Petri net AOC process model. When simulated by itself, a goal-seeking loop drives the system toward a reduction in adversary state until the joint forces commander's (JFC) goals are achieved (Figure 9.10) and all TBM observations cease. The adversary system produces its own responses by setting up and launching TBMs.

The joint forces' goal-seeking process can be halted if something causes "U.S. political will" to become sufficiently low (Figure 9.11). When simulated with the Petri net model, the goal-seeking loop is turned off, and information instead flows to and from the AOC process model.

Quantitative information can enter the system dynamics model from the Petri net through two variables: *average response time* (Figure 9.10) and *maximum personnel utilization*, which is used to determine the probability of major errors in prosecution. Figure 9.12 shows the response times for each event over the course of the

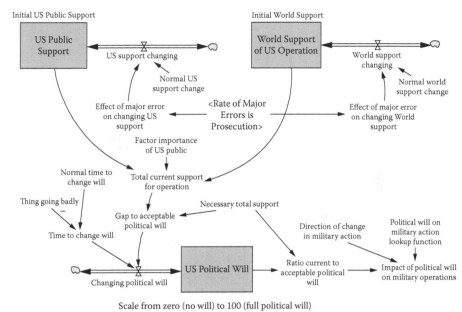

Figure 9.11   United States and world public support and its effect on political will.

day. This is the output of the AOC Petri net model at the end of day 1; average daily values for each event type are used to calibrate day 1 of the system dynamics model. The output from the system dynamics model is a set of critical events for the next day. Therefore, the two models are run in succession to simulate the effect of global dynamics on AOC operation. Some global processes may be important in influencing mission effectiveness (completing military operations to the JFC objectives). In this case, "world support" for U.S. operations, as well as "U.S. public support," both affect "U.S. political will," which decreases the degree of military actions to zero (see Figure 9.11).

9.5 Results and Discussion

Figure 9.12 shows the average event response time (*y* axis) for operators handling the two (2) last pilot-down tasks and operators handling other two (2) TST tasks for a given number of critical events. These results reflect that the pilot-down gets the top priority, thus showing a faster average response time.

Figure 9.13 shows the response times from running the AOC model for the original data file of critical events, the baseline run. More critical events in a given period result in a longer AOC response time to accomplish the tasks (e.g., see the

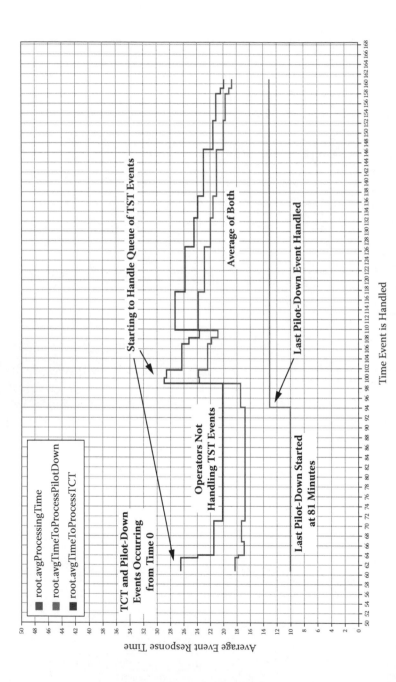

Figure 9.12 Results for average response time versus time event is handled from the agent-based model for the last pilot-down events (lower plot) and time-sensitive targeting (TST) events (upper plot); the average of both is also shown.

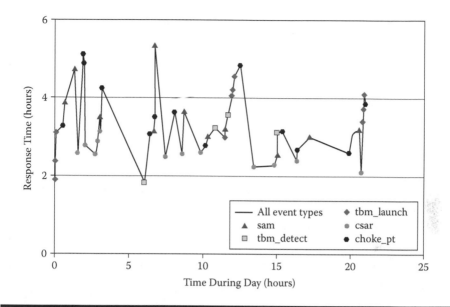

Figure 9.13 **The response time for each type of event over the course of one day (baseline run). CSAR or pilot-down has the highest priority and the remaining four events have the same lower priority.**

period between 10 and 13 hours). This indicates that operators are increasingly busy, and there is a queueing of tasks. As events become less frequent, the queues are reduced and the response time reduces (e.g., period after 13 hours). A weighted average response time was calculated for the TBM launch and detect events. This value was used in the system dynamics model: Figure 9.10, "average response time." The most-utilized operators in the AOC model were determined using the baseline run and are shown in Table 9.1. This maximum personnel utilization rate (e.g., 65%, coordinate airspace) was used in the system dynamics model.

Using the critical events in the baseline run, the AOC model was executed and data were passed to the system dynamics. The system dynamics model was run for 1 day and the output of critical events was then used to run the AOC model again. This process was done nine times and the system dynamics results are shown in Figure 9.14.

The "adversary military capability" starts at 100% and is drawn down over the 9 days. The TBM setup and launches first increase and then decrease as the capability is drawn down. United States and world public support both decrease and political will first increases slightly and then decreases. Figure 9.15 shows the critical event response time from the Petri net model for day 2 and day 9.

At the beginning of the campaign, there are more critical events and the AOC model shows that there is queueing with a maximum response time of 8.5 hours. As the critical events are reduced by taking out "adversary military capability,"

Table 9.1   Highest Utilized AOC Operator from Petri Net (Day 1)

Utilization Rate (%)	AOC Cell	Job Description
65	Coordinate airspace	Coordinate airspace
62	Combat operations evaluate current assets package mission Get approvals	Analyze event impact on ATO
61	Combat operations dynamic targeting	Determine impact; research and monitor
60	Combat operations dynamic targeting	Assess threat environment Collect intelligence Monitor emerging targets (top secret)
57	Monitor operations dynamic targeting	Determine impact Research and intelligence
55	Combat operations evaluate current assets package mission Get approvals	Asset availability Approve package for mission Analyze event impact on ATO Manage assets

the AOC model shows that queueing is reduced and response time is consistently between two and four hours. These results show that the system dynamics model has been calibrated to the Petri net model. This is the first step in model validation.

9.6   Conclusions

A preliminary agent-based model has been implemented that accurately reflects event priority: pilot-down events are handled faster than other time-sensitive targeting events. Next, the Petri net model was evaluated for use in determining information overflow indicators for operators (e.g., operator stress); implementing the attention allocation model is the first step. Operator overflow inherently has components that depend on individual characteristics. In order to gain insight into these issues, it is necessary to model individual scale dynamics. Modeling the AOC at the operational process scale and operator interactions at the individual scale are hypothesized to provide insight into operator overflow. The models and

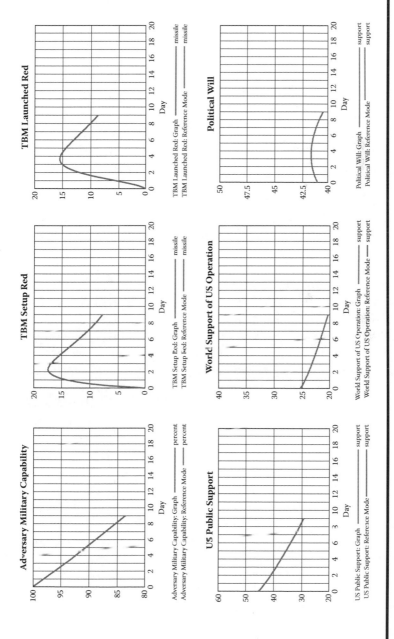

Figure 9.14 The system dynamics model output for nine days.

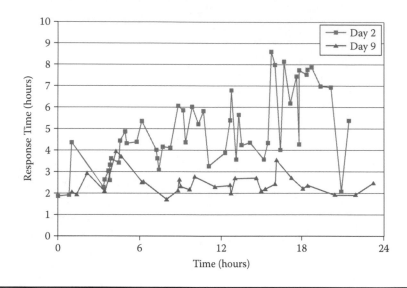

Figure 9.15   The response time for all events for the second and ninth days.

insight gained can be used by operators, decision makers, and new team members to understand the effects of policies that improve operations on the overall system.

One of the agent MOPs or indicators is errors. Future work will include modeling errors as a function of the number of activities or subtasks, the difficulty level of the subtask, and the number of interruptions multiplied by the duration. This may be an example of the agent-based model having direct feedback to the system dynamics model as was shown in Figure 9.1.

The critical events processed started at 50 on day 1 (Table 9.1) and were drawn down to 29 on day 9 in the integrated Petri net and system dynamics models. The corresponding response time and maximum personnel utilization are also shown in Table 9.2, and are reduced as expected. This drawdown scenario was selected for the first step of validation. Future scenarios could include:

- Have "adversary military capability" rise due to the purchase of weapons resulting in more critical events to handle. This would cause utilization to increase causing more errors in prosecution and possible effect on U.S. and world opinion.
- Remove a particularly well-connected operator from the AOC resulting in more utilization and higher response times.
- Currently, the CSAR or pilot-down is the highest priority event, but TBM events targeting a densely populated area could become the highest priority as dictated by the global model. This would result in faster response times for TBM, but may cause negative U.S. opinion.

Table 9.2   Average Response Time and Maximum Personnel Utilization for Each Day with Corresponding Number of Critical Events

Day	Response Time (Hours)	Maximum Personnel Utilization (%)	Number of Critical Events
1	4.33	64.5	50
2	5.40	66.1	51
3	3.40	62.0	45
4	3.54	53.9	39
5	2.79	48.8	37
6	2.76	52.3	37
7	3.09	54.5	38
8	2.34	50.3	31
9	2.37	42.7	29

The AOC process model is, however, linked (e.g., critical event process time and probability of errors) to a global-environment model that is driven by the political landscape in which the AOC operates. Future work in linking the Petri net and system dynamics models includes feedback from the global model's "priority of target" and "operators" available to the AOC; these variables will reflect changing "political will" and "JFC objectives."

References

Boiney, L.G. 2007. More than information overload: Supporting human attention allocation. Presented at 12th International Command and Control Research and Technology Symposium (CCRTS), Newport, RI. June 19–21. http://www.dodccrp.org/html3/events_12.html.

Kuras, M.L. and White, B.E. 2005. Engineering enterprises using complex-system engineering. In *INCOSE Proceedings*, Rochester, NY, July 10–15.

Kuras, M.L. and White, B.E. 2006. Complex systems engineering—Position paper: A regimen for CSE. Presented at Fourth Annual Conference on Systems Engineering Research (CSER), April 7–8, Los Angeles.

Mathieu, J., James, J., Mahoney, P., Boiney, L., Hubbard, R., and White, B. 2007a. Hybrid system dynamic, Petri net, and agent-based modeling of the Air and Space Operations Center. In INCOSE Symposium, Systems Engineering: Key to Intelligent Enterprises, June 24–28, San Diego.

Mathieu, J., Melhuish, J., James, P. Mahoney, L. Boiney, and B. White. 2007b. Multi-scale modeling of the Air and Space Operations Center. Presented at Symposium on Complex Systems Engineering, January 11–12, The Rand Corporation. http:// cs.calstatela.edu/wiki/index.php/Symposium_on_Complex_Systems_Engineering.

McCarter, B.G. and White, B.E. 2007. Collaboration/cooperation in sharing and utilizing net-centric information. Presented at Conference on Systems Engineering Research (CSER), March 14–16, Stevens Institute of Technology, Hoboken, NJ.

White, B.E. 2005a. A complementary approach to enterprise systems engineering. Presented at the National Defense Industrial Association Eighth Annual Systems Engineering Conference, San Diego, October 24–27.

White, B.E. 2005b. Engineering enterprises using complex-system engineering (CSE). Presentation to First Annual System of Systems (SoS) Engineering Conference, Johnstown, PA, June 13–14.

Acknowledgments

The authors gratefully thank James Melhuish, formerly with Aptima, Inc., for providing consultation on system dynamics modeling.

Chapter 10

Nuclear Waste Management Strategic Framework for a Large-Scale Government Program

Gregory A. Love, Christopher G. Glazner, Samuel G. Steckley, Kristin Lee, and Teresa A. Tyborowski

Contents

10.1 Introduction

This chapter presents a framework for the use of a materials flow model and dynamic influence diagram, or causal loop diagram, to develop system insight into the U.S. Department of Energy's (DoE) responsibility for environmental cleanup of legacy nuclear waste. This framework is used to explore policy options, analyze plans, address management challenges, and develop mitigation strategies for the DoE Office of Environmental Management (EM). The sociotechnical complexity of EM's mission compels the use of a qualitative approach to analysis to complement a more quantitative discrete event modeling effort. We use this analysis to drive scenarios for the model, pinpoint pressure and leverage points, and develop a shared conceptual understanding of the problem space among stakeholders. This approach affords the opportunity to discuss problems using a unified conceptual perspective and is also general enough that it applies to a broad range of capital investment/production operations problems.

In the aftermath of the Cold War, the United States was left with a formidable legacy of radioactive waste, the by-products of the creation of nuclear weapons and nuclear energy research. This is a challenging proposition, as the difficult and intensive technical process for disposition of nuclear waste is further complicated by regulatory, legal, and budget constraints. The challenge for the EM is to better understand the myriad processes, alternatives, and policy constraints of these operations from a system perspective, allowing them to better manage the system toward program completion and facility closure on time and within budget.

10.2 Background

10.2.1 Background—Mission

The DoE is responsible for cleaning up the environmental legacy from five decades of nuclear weapons development and government-sponsored nuclear energy research. In 1989, the DoE established the EM program to address these problems. Sites once involved in the production of nuclear weapons, such as the Savannah River site, are now tasked with properly disposing of surplus nuclear materials and radioactive waste by-products.

The EM mission encompasses the decontamination and decommissioning of nuclear production facilities, the safe disposal of highly radioactive liquid waste stored in underground tanks generated from reprocessing surplus used nuclear fuel

(UNF), the retrieval of nuclear contaminated waste buried at sites that are threatening the environment, and the burial of nuclear contaminated material that meets legal standards for final disposition.

10.2.2 Background—Program Management

In 1998, EM developed a "projectized" approach to cleanup, which more fully defined the life-cycle scope and cost of the EM program (DoE 1998). The Paths to Closure document marked the evolution to a more discrete project management approach for over 350 projects at DoE sites. Four years later, a comprehensive review was published (DoE 2002) recommending a renewed focus on completing projects with an appropriate sense of urgency. Program management reforms focused on performance-based contracts, comprehensive risk prioritization approaches, and business processes focused on accelerated risk reduction and tighter controls on cost and schedule growth.

In September 2005, the House and Senate Energy and Water Development Appropriations Subcommittees requested the National Academy of Public Administration (NAPA) to conduct a management review of EM. Over the course of the 19-month NAPA study (NAPA 2007), EM worked closely with the Academy panel and staff to implement recommendations during the study period before the report was published. The study panel investigated how EM was organized and managed, its human capital, acquisition, and project management operations.

Throughout this period of internal reforms, continuous improvement, and external oversight, EM has been evolving its management practices and business systems. EM has formalized these efforts with "Journey to Excellence" initiatives to institutionalize the evolution to best-in-class processes and practices.

10.2.3 Objectives

The EM program scope illustrates the complex system of systems inherent in large-scale government programs. The program spans a long time interval, with completion estimates extending out to the 2050 and 2062 timeframe (DoE 2010, p. 8). Large investments in the billions of dollars are involved. The risks are very high. Cost escalation, delays, and technical problems can undermine the financial feasibility, jeopardize its completion, and lead to government inquiries. Problems in any single dimension can pose substantial management challenges. The challenge for EM is to better understand the myriad processes, alternatives, and policy constraints of these operations from a systems perspective, allowing them to better manage the system toward program completion and facility closure on time and within budget while meeting performance measures.

To date, DoE has reduced the sites requiring cleanup from 110 to 18, which represents a reduction in the legacy footprint from 3,125 square miles to 900 square miles (DoE 2011). Despite this progress, the remaining work presents unique management, technical, and stakeholder challenges. Within this mission, the chief threat to the environment, health, and safety is the radioactive liquid waste. The DoE currently manages approximately 88 million gallons of highly radioactive waste in 239 underground tanks. Collectively, these tanks and downstream operations are the largest cost element in the EM program (DoE 2011).

The EM program prioritizes (DoE 2010, p. 5) activities that are projected to reduce the most curies per volume (the curie is a unit of radioactivity). These activities include (but are not limited to):

- The treatment and disposal of liquid waste stored in underground tanks
- The receipt, storage, and disposition of UNF
- The consolidation, stabilization, and disposition of special nuclear materials

This chapter presents the influence diagrams and the model structures that are currently being applied to address these objectives.

10.2.4 A Unifying Structure

There is a large body of work on the application of system dynamics to project management. Lyneis and Ford (2007) have surveyed published literature with a focus on single projects. Very large-scale capital projects in the public sector have been singularly analyzed in a case study format (Lyneis, Cooper, and Els 2001). We have applied these and other causal structures to identify scenarios that the model might explore. Although qualitative in nature, the influence diagrams capture the relationships between key variables and formalize the mental models of decision makers and engineers.

This chapter presents a framework for exploring policy options, analyzing plans, addressing management challenges, and developing mitigation strategies. This framework makes it possible to see a complex problem on a single sheet of paper and affords the opportunity to discuss problems using a unified conceptual perspective. The framework is also general enough that it applies to a broad range of capital investment/production operations problems. The causal influences also identify those feedback loops that represent significant management challenges to the DoE and can be generalized to large-scale operations in both the public and private sectors.

What sets this framework apart from previous work is the system of systems scale and the joint operations/capital project dependencies and complexities. The production planning and operations of existing physical facilities need to be accomplished in an efficient, timely, and cost-effective manner. The physical characteristics of surplus nuclear materials stocks and radioactive waste streams are dynamic and often require investments in new technologies for safe disposal.

Project outcomes need to be viewed in the context of their impact on ongoing and future operations. Today's decisions have to be evaluated in the context of a common framework that can be translated into a model to generate reliable performance measures and outcomes. Collectively, these structures are combined into an influence diagram (Coyle 1996, 2004). The diagram identifies the key variables and policies that are of particular interest to the sponsor (EM). At a more technical level it identifies the main features of the problem addressed by the model. We expand the influence diagram to illustrate generic operational structures at a key government site. Although the modeling activities are ongoing, the chapter highlights insights gained from early results.

10.3 Development of the Influence Diagram

Qualitative diagrammatic modeling in the form of influence diagrams is used to communicate the model scope and describe the relationship between key variables in the model. Many of the key variables are explicitly modeled, however, some emerge from the scenario analysis. The influence diagram is an overall system representation that can be used to design scenarios.

10.3.1 Physical Flow—The Route to Closure

One of EM's goals is to accelerate the cleanup and reduce the life-cycle costs of legacy materials. To achieve this goal, DoE uses legacy-hardened production facilities to reprocess used nuclear fuel, and separate and treat waste products. There are cases, however, where new capabilities are required to treat the radioactive wastes and prepare them for final disposition. Figure 10.1 shows the investment and production chain associated with the transformation process. EM is responsible for the disposition of surplus nuclear material stocks and nonproliferation

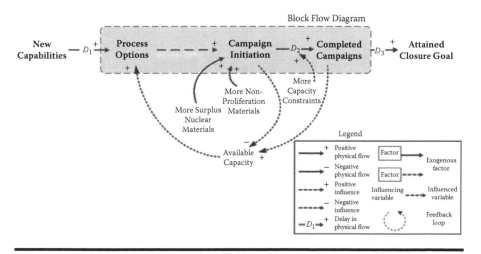

Figure 10.1   Capital investment and production campaigns.

stocks. Management allocates production resources by scheduling campaigns for these different materials. Each campaign has a distinct start and finish date and is organized into a master roadmap. With each campaign start, existing production capacity is committed for that purpose, temporarily reducing the available capacity for other campaigns. Capacity is subsequently freed when a campaign ends, making process options available for other materials. This flexibility is shown by the influence from available capacity to process option in Figure 10.1. Over the course of the campaigns, interim milestones mark periodic progress toward a final closure objective.

This planning process is straightforward for conventional materials that can be processed in existing facilities. However, unconventional materials often require capital investments with first-of-a-kind technologies. These investments can range from minor modifications to a major investment in a new facility such as the Salt Waste Processing Facility (SWPF).

Investment decisions are generally driven by production schedules and stakeholder commitments. This is more typical in the public sector in contrast to the private sector. Morecroft (2007) presents three different approaches to evaluating capital investments: finance-driven, planning-driven, and operations-driven. EM typically focuses on the required capacity to meet regulatory commitment dates and projected benefits from accelerating milestones (operations-driven). This capital investment approach is a viable rationale; we simply point out that it is generally more appropriate for the public sector. EM manages these investments to deliver performance objectives on time and within budget. Over the past five years, there has been a focus on accelerating the cleanup by compressing the roadmap plan and generally managing total program costs to a level funding profile.

10.3.2 Production Complex Block Flow Diagram

The available process options are a reference to the production facilities and infrastructure represented in a block flow diagram. The generic block flow diagram in Figure 10.2 identifies the facilities at a single location and is similar to the types of models described by Forrester in *Industrial Dynamics* (1961).

10.3.3 Budget and Funding Levels

The cost components add an important strategic context to the model. The ability to derive a total life-cycle cost makes it possible to monetize resources (labor, production assets, and investment) for any scenario. This enables management to take corrective action based on simulated cost profiles.

Figure 10.3 shows the funding policy decision and the process of allocating funds to operations and investment. The aggregate budget is primarily set by exogenous funding decisions. Annual appropriation bills establish the program budget. EM management can exercise some discretion to allocate expenditures between

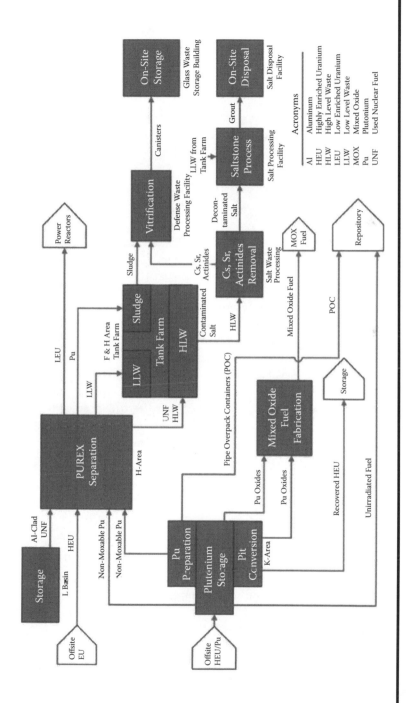

Figure 10.2 Production block flow diagram.

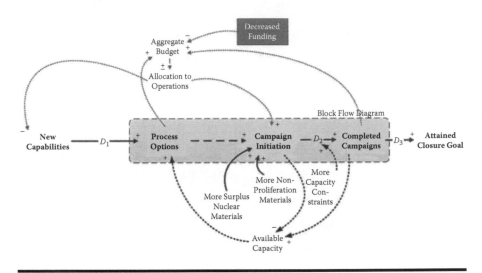

Figure 10.3 Sources and uses of funds.

operations and investments. This provides leverage to accelerate prioritized closure activities to meet critical program objectives.

Production campaigns consume resources that are monetized in the model. A large proportion of these costs are direct labor operating expenses. However, there are incremental activity-based costs tied to discrete operations. When the campaigns are completed, funds become available for other purposes. There is a parallel structure for investments. When major construction activities are completed, construction funding winds down, freeing resources for other activities. This should not be interpreted to mean that prior funding levels could be reallocated for other purposes. The funding policy usually restricts gross reallocation, but it may enable the capital project to transition to an operating phase. This is modeled as a state transition from an investment to an operating facility.

10.3.4 Policy Influences

The cost estimate for cleaning up the radioactive tank wastes is between $88 billion and $117 billion over the next 40 to 50 years (DoE 2010). With a planning horizon this long, there will be opportunities to accelerate tank closures with investments in new technologies and strategic operating decisions. The Accelerated Closure Policy (DoE 2010) reflects this posture, making investments in new capabilities and increasing the surplus nuclear materials production rate to accelerate the closure date.

A proactive nonproliferation policy would have a similar effect, the main difference being the introduction of more nonproliferation materials from outside the DoE complex. New investments and more campaigns may be required to treat nonproliferation materials. These policy influences are diagrammed in Figure 10.4.

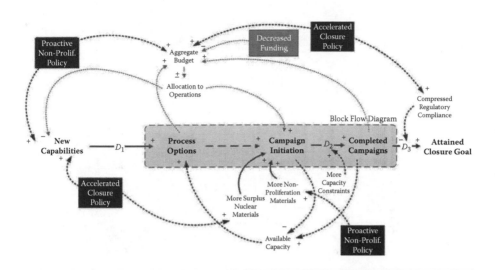

Figure 10.4 Policy influences from closure and nonproliferation decisions.

10.3.5 Needs and Project Outcomes

As policy decisions increase the stocks of surplus nuclear materials and nonprolifer-ation materials, the need for new capabilities creates a capability gap that becomes the justification for new investments. There are two feedback loops associated with capability gaps. The first is a reinforcing loop, R_1: demand for additional capac-ity. As more production campaigns are initiated, these activities tie up the pro-cess equipment, increasing the capability gap and the need for new capabilities. New investments in production capacity may be required to eliminate the capacity shortfall. The implication is that life-cycle acceleration may become capacity con-strained in the absence of new investment.

A relief strategy can be seen in the balancing loop, B_1: early completion mitigates capacity constraints. Completing campaigns frees up capacity, closes the capability gap and may obviate the need to expand capacity. The challenge is to develop a life-cycle campaign strategy that strikes a balance between these two feedbacks in such a way they minimize investment and maximize production flexibility. The underly-ing model is designed to explore this trade space. This approach is consistent with recommendations to prioritize cleanup work to achieve the greatest technical risk reduction at an accelerated rate (DoE 2002, p. II-3).

Project outcomes effect new capabilities as in the reinforcing loop R_2: schedule slip exacerbates the capability gap and illustrates how delays in the delivery of new capabilities prolong the capability gap, putting pressure to resolve the problem with stopgap measures and acceleration strategies. Figure 10.5 only illustrates the effects of schedule slip; a similar reinforcing loop for cost and performance outcomes can cause a project to spin out of control. For example, a performance shortfall can also fail to narrow the capability gap and in the worst case could require a follow-on

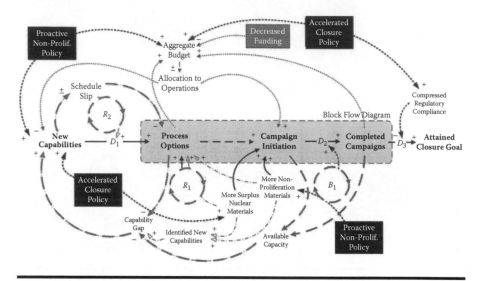

Figure 10.5   Identifying and delivering new capabilities.

project to address the deficiency. The diagram exposes the life-cycle consequences of large-scale projects that fail to deliver in any combination of the three outcomes: cost, schedule, and performance.

Feedback loops:

R_1: Demand for additional capacity.
R_2: Schedule slip exacerbates the capability gap.
B_1: Early completion mitigates capacity constraints.

10.3.6 Stakeholder Engagement

EM works with the congress, regulators, stakeholders, and tribal nations to fulfill requirements under existing regulatory agreements and comply with current environmental laws and regulations. This engagement is important to the DoE in order to efficiently accelerate risk reduction strategies as opportunities are identified.

Figure 10.6 illustrates a series of reinforcing loops, R_3: Timely progress increases support for new starts. With successful campaign completions, stakeholders are more likely to approve requests for construction (denoted by the start of new capabilities) and operating permits (receipt of new materials leading to the start of campaign initiation). Timely progress on existing commitments will increase the likelihood that stakeholders will support new starts.

Feedback loops:

R_3: Timely progress increases support for new starts.
R_4: Timely closures speeds regulatory approvals.

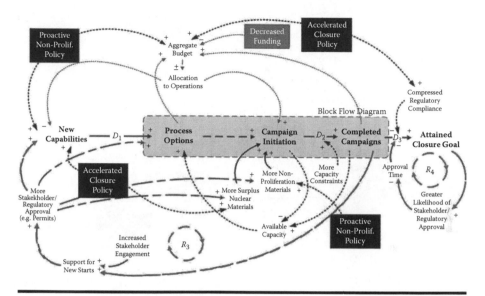

Figure 10.6 Addressing stakeholder concerns.

In a similar fashion, emptying and closing waste tanks on schedule will increase the likelihood that regulators will formally approve tank closures and accelerate the final step in the closure process. This behavior can be seen in R_4: Timely closure speeds regulatory approvals.

10.3.7 Complete Influence Diagram

The stepwise building of the influence diagram introduces the problem complexities systematically and logically through a gradual process that effectively captures the causes of dynamics. Each step focuses on a different dimension. By initially breaking down the problem and then reconstructing the dynamics iteratively, a series of individual mental models is honed into a more complex series of system interactions that establish a level of understanding that sharpens initial perceptions. The resulting diagram effectively captures the collective understanding of the team (Sterman 2000). Figure 10.7 illustrates the whole influence scenario in a single diagram. While the diagram may seem to be too broad-brush, each of the iterative builds can be disaggregated to expose more detail. Several opportunities for these excursions were previously identified. In fact, this diagram was used as the conceptual model for developing a fully fledged dynamic simulation model.

The diagram has been used as a starting point to identify and explore scenarios for the simulation model. These scenarios evaluate more specific hypotheses that are subsequently developed in the model. Many of the scenarios are "what if" experiments that explore the consequences associated with the timing of certain key decisions and events. Although the influence diagram may appear to be too general to

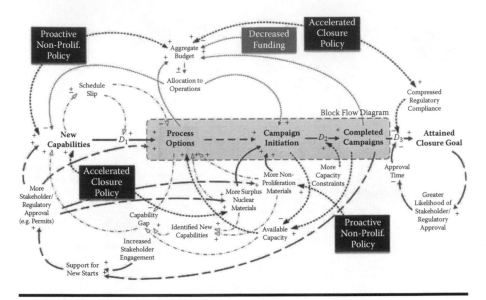

Figure 10.7 Complete influence diagram.

address feedback loops, leverage points, and more complex system integrations at the operational level, the simulation model permits more detailed investigations. Experience has shown the benefits from summarizing the results by referring back to the high-level interactions in the influence diagram.

The scenarios have led to a deeper understanding of systems dependencies between closely coupled operations. New capabilities that accelerate the waste cleanup have uncovered technical challenges with designing a robust system that can smoothly transition to the new operating state during the initial startup phase and continue during steady-state operations. For example, a new production unit introduces new interactions and systems dependencies with existing capabilities that may result in cascading effects during system upsets. The model has helped identify these circumstances and develop mitigation strategies.

10.4 Conclusion

The influence diagram in Figure 10.7 is an overall system representation in a single diagram. The level of aggregation masks some of the details; however, the advantage lies in the ability to analyze the problem from a high-level systems perspective. By probing the relationships between key model variables, the diagram effectively conveys the problem complexities. The stepwise progression through the diagram hones the collective mental models into a more cohesive whole and leads to a deeper understanding.

The analysis of feedback loops promotes the development of scenarios that can be evaluated in more detail with a simulation model. These model runs may test important subsystems, explore system resilience, identify leverage points, or develop system plans that satisfy life-cycle criteria. The results of these runs can then be generalized in the context of illustrative planning scenarios using the influence diagram to summarize important findings.

References

Coyle, R.G. 1996. *System Dynamics Modelling: A Practical Approach.* London: Chapman and Hall/CRC Press.

Coyle, R.G. 2004. *Practical Strategy: Structured Tools and Techniques.* Harlow, UK: Pearson Education, pp. 29–46.

DoE (U.S. Department of Energy), Office of Environmental Management. 1998. *Accelerating Cleanup: Paths to Closure* (June). http://www.em.doe.gov:Publications:accpath.aspx.

DoE (U.S. Department of Energy), Office of Environmental Management. 2002. *Top-to-Bottom Review of the EM Program* (4 February). http://www.em.doe.gov/pdfs/16859ttbr.pdf.

DoE (U.S. Department of Energy), Office of Environmental Management. 2010. *Roadmap for EM's Journey to Excellence* (16 December).

DoE (U.S. Department of Energy). 2011. *Draft Strategic Plan* (February): 40–42.

Forrester, J.W. 1961. *Industrial Dynamics.* Cambridge, MA: MIT Press.

Lyneis, J.M. and Ford, D.N. 2007. System dynamics applied to project management: A survey, assessment, and directions for future research. *System Dynamics Review,* 23(2/3): 157–189.

Lyneis, J.M., Cooper, K.G., and Els, S.A. 2001. Strategic management of complex projects: A case study using system dynamics. *System Dynamics Review,* 17(3): 237–260.

Morecroft, J. 2007. *Strategic Modelling and Business Dynamics.* Chichester, UK: John Wiley & Sons, pp. 204–205.

NAPA (National Academy of Public Administration). 2007. *Office of Environmental Management: Managing America's Defense Nuclear Waste* (December). NAPA:7-15. http://www.napawash.org/wp-content/uploads/2007/07-15.pdf.

Sterman, J.D. 2000. *Business Dynamics: Systems Thinking and Modeling for a Complex World.* Boston: Irwin McGraw-Hill, p. 137.

Chapter 11

International Trade and Commerce: Enterprise Systems Engineering and Architecture in a Multiagency Environment*

William J. Bunting and Kenneth C. Hoffman

Contents

---

* Adapted, in part, from K.C. Hoffman et al., Enterprise business, computing, and information services in a multiagency environment: A case study in enterprise architect-engineering, In International Enterprise Distributed Object Computing Conference (EDOC) Workshop, IEEE Computer Society. © 2005 IEEE.

11.1 Introduction

This case study formulates and demonstrates a comprehensive planning framework for enterprise systems engineering (ESE), an integrated ESE workbench for use by multinational and multiagency stakeholders, public and private, engaged in international commerce. The objective is to evaluate alternative strategies and solutions, and to support program management in the acquisition and implementation of the selected solutions and systems by individual entities that contribute to the highly integrated global trade system.

The modernization of logistics management with emergent computing, communication, and information services is central to increased performance capabilities in a networked enterprise such as international commerce, and a high-risk venture in any environment. Fully aligning these services with the high-performance operational objectives of enterprises has proven to be difficult. To deliver optimal products and services and high-quality business outcomes for both internal and external stakeholders, enterprises are expected to have information environments that fully integrate business processes and systems.

Many of the challenges faced by modern society—international trade, healthcare, and national security—require complex multiagency responses involving public and private networked institutions cooperating on a global scale. The effective integration of these institutions' information systems with modernized business processes is an absolute requirement and represents a major engineering and governance challenge.

The case study was performed in a research program sponsored by The MITRE Corporation. It applies enterprise architecture and engineering tools and methods to monitor and manage international trade in an environment of heightened security risk, described by Hoffman et al. (2005). Simulation models, web services, and semantic technology are used as "enterprise middleware" to integrate enterprise architectures with geographic information system-based business performance planning and analysis tools into the integrated ESE workbench. Full alignment of enterprise information services with business processes is achieved by applying governance principles of activity-based management and enterprise resource planning. Additional decision methods and theories will be required to deal with situation-specific complexities and uncertainties faced by stakeholders.

11.2 Description of the Enterprise— International Commerce

The significant undertaking, the enterprise, in this case study is the management of international commerce activities and improvements in the effectiveness and security of trade to support the global economy. This international trade enterprise is composed of private-sector and public entities that manage and monitor trade including suppliers, transporters, and purchasers operating the supply chain in conformance with government policies and procedures at ports of entry.

11.3 Statement of the Enterprise Transformation Challenge

The transformation envisioned for this case study is the development and implementation of secure multinational trade technologies and systems that are utilized for commercial transactions and tracking by purchasers, sellers, and logistics managers (private sector) as well as government programs initiated for customs operations and security purposes.

The individual firms and agencies engaged in international commerce make decisions and investments that drive the transformation to improved performance levels. Decisions and actions are sometimes coordinated through government initiatives, but often are taken independently for local reasons adding to the complexity. The case study postulates a "notional" joint program office operating in the public interest to monitor and analyze individual initiatives, integrate the results, and project the likely outcomes of the collective efforts to inform decisions and actions taken by all stakeholders. No such office exists although the potential benefits are significant.

Management of the supply chain for production relies on the successful implementation of real-time information systems and sensors to track and locate commercial products, and to record transactions at points of sale, ports of entry, and delivery. Information-sharing technologies must provide accurate and timely information on shipments and their provenance to comply with the advanced security and privacy provisions required by governments.

An ESE perspective is required to characterize and analyze the integrated technologies and organizations that lead to excellence in planning, transforming, and operating the international trade system. Programs to modernize computing, communication, and information services are high-risk ventures with very low success rates. It has proven especially difficult to fully align these services with new business processes that are often deployed at the same time. The enterprise transformation that is required involves all elements of an enterprise: management, organizations, technology, and information. Further complicating modernization efforts are the rapid pace of technology advancement and the emergent behavior it facilitates inside and outside the enterprise. Increasingly, the performance required of enterprise information systems is fully known only in retrospect. Hence, the enterprise architect needs to plan for uncertainly and build in agility. While organizations are "groping and coping" with these challenges, the increased demand for multiagency services is raising risks to an even higher level.

The complexities of international trade as a system include the large and distributed scale of operations; multiagent roles and responsibilities and associated governance challenges; technical and system diversity of technologies, policies, and management methods; emergent behaviors; security concerns; and major uncertainties raised by ever-changing sources and markets. Milly et al. (2008) states that "stationarity is dead." His statement refers to water management as usage increases to levels where availability is affected by the always present vagaries of weather and competing demands, but applies in an even stronger way to the very wide range of factors to be addressed by ESE. Many systems engineering efforts assume a level of "stationarity" to simplify the problem, an assumption that cannot generally be made for ESE challenges such as the international trade system.

11.4 Characterization of the International Commerce Enterprise

The central features of international commerce are presented in a geographic perspective on the flow of goods in a supply chain from their source to destination point of sale. The geographic characterization draws on public data sources in the form of a geographic information system (GIS), Figure 11.1, which describes the major elements of a demonstration exercise and the activities involved in international commerce. The description is intentionally at a generic level so that it applies to any geographic region engaged in secure global commerce. The case study exercise serves to develop and apply an ESE (or architect-engineering) approach to international commerce, a sector of major economic importance that poses serious security concerns. Specific activities and related information services defined include the following:

- Cargo import and export processing (e.g., account maintenance, data mining, and inspection services)
- Passenger security processing at departure and arrival (e.g., image processing, data mining, and decision support services)
- Apprehension and identification of individuals (e.g., biometric, data mining, and communication services)
- Financial flows in international commerce (e.g., workflow, data mining, and decision support services).

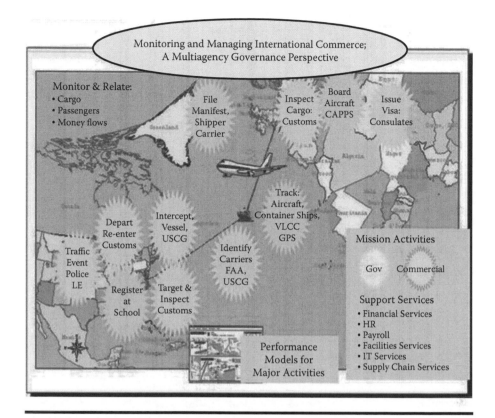

Figure 11.1 ESE characterization defining the end-to-end activity sequence for a multiagency mission in international commerce, including the monitoring of all governmental and commercial aspects of global commerce, and indicating two possible "apprehension events." (Reprinted from K.C. Hoffman, T. Pawlowski, D. Payne, and K. Zheng, in International Enterprise Distributed Object Computing Conference (EDOC) Workshop, IEEE Computer Society. © IEEE 2005. With permission.)

Each activity is the responsibility of an individual public agency or private entity and must be coordinated with effective information sharing (within security and privacy constraints) to meet the overall mission objective of furthering economic growth while protecting nations from the illegal entry of people and goods. The elements of the enterprise provide a complete description of the activities performed by participants in a multiagency mission. The GIS can be used to illustrate the end-to-end activities involved in cargo and passenger flow between continents.

State-space descriptors of the international trade enterprise include:

- Source-destination patterns and modal mix
- Quantities of goods delivered (physical and monetary terms)

- Supply chain inventories
- Time, total source to destination, entry processing
- Organization of enterprise: number of firms, levels of integration or consolidation
- Cost and price information and projections for investment analysis
- Quantities of labor, energy, and other resources required at each step in the supply chain

The supply chain is controlled directly by exporters, transporters, and importers using logistics management methods. Bar codes and radio frequency identification (RFID) technologies are ubiquitous on containers and products, and enable tracking of shipments from source to destination.

Control parameters that affect the evolution of the enterprise state include:

- Levels of investment in carriers, sensor technologies, and supply chain systems and software
- Organizational changes
- Policy changes in supplier and consumer regions that affect trade patterns and investments

Policies governing the entry of goods are controlled by national governments to collect appropriate tariffs and inspect food and agricultural products, subject to additional local regulations regarding the transport of hazardous materials. Governments also regulate trade in critical technologies with national security implications. Major risks and uncertainties that both industry and government must cope with are accidents, disasters, theft, and terrorism.

11.5 Strategies and Technologies Examined for the Transformation

The transformation strategy is to develop and implement network-centric information management and communication technologies for the multinational trade system. The network concept encompasses the organizations that must interact in real-time, as well as the data management and communication technologies that provide situational awareness. The case study provides a proof-of-concept for the planning and management of the enterprise transformation program and commercial trade operations.

Multiagency missions are extremely demanding from both the organizational and technical perspectives. They require the integrated efforts of executive managers, operations managers, and engineering managers. This is especially critical in the case of highly networked multiagency operations—the net-centric enterprise—in which both organizational and policy coordination and technical interoperability of information and communication services are required. Specific planning and management functions that can be integrated and that benefit from the scope and content of the ESE workbench include the following:

- Strategic planning
- Performance planning
- Operations planning and management
- Engineering management
- Information systems planning and management
- Training

The ESE workbench supports the objective of the "notional" multiagency governance environment that builds on the capabilities of individual agencies for a close coupling of strategic planning, technical management, and operational management. The workbench approach, with a strong focus on mission activities, can deal effectively with complexity and challenges in achieving this objective. It provides the comprehensive systems perspective of the modernization technologies and mission activities necessary for successful program coordination and management in a multiagency environment.

Many alternative management options are available for both modernization programs and operations. Regardless of the option selected, achieving successful multiagency governance requires a robust governance structure operating on a comprehensive information base. An architecture-based approach to governance can help resolve complexity and the many organizational accountability and control issues that arise. This approach ensures that routine management tasks are integrated across multiple agencies and are responsive to strategic direction and assigned responsibilities.

Defining the required level of organizational and technical interoperability is a major factor in multiagency missions. The level of interoperability, which depends on the specific mission, can vary among the specific activities performed and is always a strong factor in the assignment of responsibilities and accountability. Specifically, the level of information sharing and provisions for accountability, information quality, privacy, and security are tied closely to the level of organizational and technical interoperability required.

11.6 ESE Analytical Approach

A number of management and engineering tools and methods have been developed and employed for specific aspects of enterprise planning and analysis. These were surveyed and a set of tools was selected to comprise the integrated ESE workbench to address the complex multiagency operational environment in international commerce. The workbench implements a web-based enterprise planning and management approach that enables decision makers to visually plan future states of the enterprise and operations performance. With this integrated planning tool set and web-enabled capabilities, one can readily drill down to the lower-level details of principal elements of the ESE workbench, including the following:

■ Enterprise architectures: For information on mission operations, the organization, activities, processes, services, and component technologies for the enterprise
■ Business process simulation models: Discrete event models for dynamic representation and measurement of business operations, supporting services, and resources required to perform the enterprise mission or deliver services to citizens
■ Geographic information systems (GIS): For portrayal of the spatial, environmental, and demographic dimensions of the operations
■ Common repository of planning information: To ensure consistent information content across the tool set and to integrate tools and methods using web services and semantic technologies

The workbench enables a line of sight (LOS) approach to performance planning in which the roles of all business and information assets and services, existing and proposed, are described and analyzed in the context of simulated mission operations. The value of specific technologies and services may be gauged with a line of sight to analyze how effectively they support mission performance following the approach developed by Bunting (2005). The core elements of the ESE workbench are illustrated in Figure 11.2.

These seven elements are applied in the following sequence:

1. *Operations plans and scenarios:* The initial plan and operations scenario for the multiagency mission provides the focus and content for the simulation and architecting steps to follow. The plans and scenarios are updated and improved as a result of the analyses performed using other elements of the ESE workbench, particularly performance analysis capabilities.
2. *Dynamic operations and systems performance simulation models:* Process models are employed to simulate the performance of business operations using the precise business activities represented in the operations-centric architecture (see Figure 11.3). Physical flows in the logistics supply chain, and processing at ports of entry, can be modeled for performance analysis, along with supporting information services using computing and communications system tools such as capacity planning, network simulation, and data management. Performance simulations at port of entry have been developed by Payne, Hoffman, and Zheng (2003) to analyze the cargo processing dynamics including staffing and equipment requirements to manage entry traffic. Linkages between information services and business process performance are defined through operating level agreement (OLA) and service level agreement (SLA) specifications that serve as requirements statements for selecting

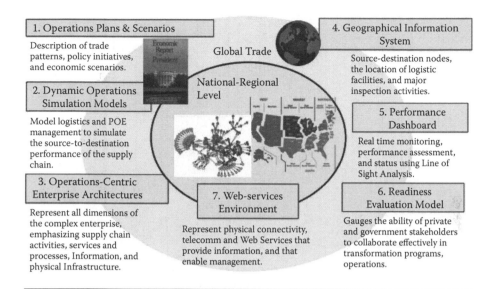

1. Operations Plans & Scenarios

Description of trade patterns, policy initiatives, and economic scenarios.

2. Dynamic Operations Simulation Models

Model logistics and POE management to simulate the source-to-destination performance of the supply chain.

3. Operations-Centric Enterprise Architectures

Represent all dimensions of the complex enterprise, emphasizing supply chain activities, services and processes, Information, and physical Infrastructure.

Global Trade

National-Regional Level

7. Web-services Environment

Represent physical connectivity, telecomm and Web Services that provide information, and that enable management.

4. Geographical Information System

Source-destination nodes, the location of logistic facilities, and major inspection activities.

5. Performance Dashboard

Real time monitoring, performance assessment, and status using Line of Sight Analysis.

6. Readiness Evaluation Model

Gauges the ability of private and government stakeholders to collaborate effectively in transformation programs, operations.

Figure 11.2 Elements of the integrated enterprise systems engineering (ESE) workbench for transformation planning and management of supply chain operations in a complex multiagent private, state, local, and federal environment. (Reprinted from K.C. Hoffman, T. Pawlowski, D. Payne, and K. Zheng, in International Enterprise Distributed Object Computing Conference (EDOC) Workshop, IEEE Computer Society. © IEEE 2005. With permission.)

systems and components, or managed commercial services. Information services are modeled using capacity planning tools for distributed computing along with network design tools for communications and other conventional information systems design and simulation methods. All mission activities, resources, and assets are described. Explicit relationships are established here with the architectures of the agencies, bureaus, or commercial entities that support the mission activities as described in the case study.

3. *Operations-centric enterprise architecture:* The operations-centric architecture, Figure 11.3, integrates applicable elements of the enterprise architectures and information systems architectures of individual agencies (Zachman 1987). The enterprise levels, business activities, services, data, and technology, are defined in various architecture frameworks promulgated by the federal government (CIO Council 2001) and the U.S. Department of Defense (DoD 2003). The more recent Office of Management and Budget Federal Architectural Model (OMB-FEA 2003) provides descriptive elements encompassing all aspects of an enterprise: performance, business, services and components, data and information, and technology (infrastructure). The operations-centric architecture uses these same categories, and case study results are discussed for each. Integration of the individual enterprise architectures into the operations-centric architecture is driven by the linked activities conducted

Operations-Centric Architecture

• Represents **Activities** performed in the operations, and **Information Services** utilized

• Includes **data and infrastructure** selected from agency resources

Extract Applicable Activities & Services

Select Data and Technical Resources

Identify Gaps:
• *Activities*
• *Services*
• *Policies & Plans*
• *Data*
• *Infrastructure*

Agency Enterprise Architectures

• **Activities and Services** have semantic "tags"

• **Data and Infrastructure Technology** associated with **Services**

• Use different **Frameworks with Reference Models**

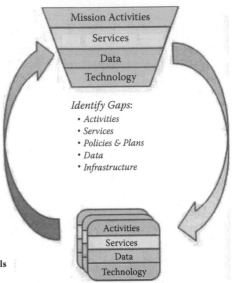

Figure 11.3 Operations-centric architecture, the central element of the ESE workbench. (Reprinted from K.C. Hoffman, T. Pawlowski, D. Payne, and K. Zheng, in International Enterprise Distributed Object Computing Conference (EDOC) Workshop, IEEE Computer Society. © IEEE 2005. With permission.)

and the information services employed by multiple agencies to perform the mission. This architecting approach relates agency program plans represented in their architectures to the overall mission effectiveness and operational excellence.

4. *Geographic information systems:* Trade operations are performed on a widely distributed, sometimes global, scale. Accordingly, the geographic aspects of operations, the supply chain, service delivery, and resource location are of great importance. The use of GISs enables the portrayal of all physical assets and enterprise resources for the multiagency operation and supports their intelligent allocation.

5. *Performance dashboard:* The performance dashboard portrays all geographic and organizational aspects of multiagency operations. It presents the performance results of simulations to identify strengths and weaknesses in technical and governance capabilities as well as gaps or overlapping organizational responsibilities and information services.

6. *Readiness evaluation model:* The readiness evaluation model is a structured method developed by Sowell (2005) for defining and evaluating the readiness of agency and private sector organizations participating in multiagency operations to interact at the level required for performance. Five levels are

defined, ranging from the least stringent requirement that agencies merely be aware of one another's activities in an operational region or with respect to a citizen service, to the most extreme case in which joint management and a central point of operational control are required.

7. *Web services environment—"enterprise middleware" and repository:* It is important that consistent definitions be used to share appropriate information on a timely basis among all elements of the ESE workbench. This is accomplished using a common repository for planning information along with web services technology, which enables aggregation and rebranding of heterogeneous services across the enterprise. Web services technology provides a single common framework for business services and mission partners.

Other standards applied in the workbench include business process executable language (BPEL) for web services and semantic web services using the resource description framework (RDF) and web ontology language. Semantic technologies are important to application integration solutions because they provide a shared understanding of data and achieve knowledge sharing by formalizing the application semantics among multiple organizations and agencies. The use of BPEL offers additional capabilities for the ESE workbench to integrate processes and businesses based on a standard format, metadata, and a common set of notations (Payne, Hoffman, and Zheng 2003).

Application of a consistent set of mission/business activity and information service definitions across the set of tools in a form that can be embedded in enterprise resource planning (ERP) and other management systems, positions the ESE workbench as a major analytical asset for supporting enterprise governance. Specific functions of enterprise governance that are supported are noted in Figure 11.1.

11.7 Governance

These governance recommendations focus on the "notional" joint program office to coordinate public and private acquisition and management activities during the transformation. Specific governance roles and responsibilities are summarized in Table 11.1 using the architecture format as a guide. Authorities and accountability must also be delegated to complete the roles, responsibilities, authorities, and accountabilities (R2A2) approach. The concept of managed services is central to this concept, where business and information services are related directly to mission activities. Specific service performance measures and metrics are captured in SLAs and OLAs. These agreements are equally important whether services are provided by internal agency resources or through commercial contracts.

Table 11.1 Summary of Governance Roles and Responsibilities for a Multiagency Mission Program Office

Architectural Hierarchy	Governance Role	Responsible Entity
Performance (OMB-FEA)	Performance Management Performance Planning Performance Measurement Modernization Management	Joint Mission Director Strategic Planner Financial Manager (Agency CFOs) Technology Manager
Mission, Business	Business/Mission Management Business/Mission Planning: • Strategic • Tactical Enterprise Architecture Management EA Planning	Mission Manager (Agency Activity Managers) Strategic Planner (Agency Plans) Command Center Manager Chief Mission Architect Information Services Planner (Agency CIOs)
Services, Components	Managed Business and Information Services: • Information Services Management • Modernization Program Management	Service Delivery Coordinator (Agency service managers) IT Modernization Coordinator (Agency CTOs)
Data and Information	Managed Data Services: Database Administration	Data and Information Coordinator (Data service providers)
Technology	Managed Utility Services: Infrastructure Utility Management	Infrastructure Services Manager

The importance of governance in major programs is stressed by Hamaker (2003), whether managed internally or contracted to external parties as managed services. Well-defined roles and responsibilities facilitate support functions through:

■ Strong accountability for mission activities performed under central direction
■ A virtual support structure with central mission integration, but distributed support resources and information

The application of the mission-centric architecture is to determine responsibilities and to manage accountability. Responsibilities are matched to levels of the architecture to assure integration and coordination under the managed services approach as shown in Table 11.1. The responsibilities of supporting agencies are noted in parentheses. For the roles and functions indicated, the responsible entities are the principal managers and users; they are also accountable for their currency of planning materials. These capabilities do not come easily. All of the architectural elements are the subject of detailed systems engineering analysis and design. Architectures can play an important role in governance as described here, but require specific additions to their scope and content as dictated by the specific multiagency mission.

Finally, the utilization of architectures in integrated governance of mission operations requires a higher level of specificity, as well as currency, of the architecture products and artifacts to represent the immediate situation. The process also requires ready access to the contents of architectures and plans, a significant technical challenge currently under investigation through such advanced techniques as semantic web services using eXtensible Markup Language (XML) and similar technologies.

11.8 Results and Recommendations

Results of the more detailed activity and process modeling conducted for selected business processes are as follows:

- Enterprise architectures constructed at the single-agency level focus primarily on the technical infrastructure, and gaps exist in the mission and business activity descriptions. The operations-centric architecture described here uses the activities simulated in the process models and identifies specific activities to be added at a consistent level of detail across participating agencies.
- Critical organizational interactions are identified, along with overlaps in roles and responsibilities among participating agencies conducting a specific activity. These can be resolved in the operations-centric organizational view.
- Duplicative information services employed to perform similar activities across agencies are identified. The performance simulation models enable a comparative analysis of duplicative services, which can be used to select a service provider for an activity and enable measurement of the improvement resulting from the introduction of technology.

The case study examines performance issues in selected activities of the mission operations incorporated in the operations-centric architecture and performance simulation. Representative data from public sources were used to provide a realistic ESE workbench proof of concept. The analytical results do not apply to any specific nation, agency, or location. Examples of general findings and results of applying the workbench to all aspects of the enterprise scope as outlined in the OMB-FEA reference models include the following.

11.8.1 Performance Analysis and Management

The modernization vision of this multiagency mission is illustrated as a collaborative environment for monitoring all aspects of international commerce involving imports and exports between the United States and Europe. This includes government systems as well as private-sector supply chain management systems. The individual process simulation models, linked in an end-to-end cargo and passenger flow sequence, represent the throughput and cost performance at each step as well as the overall flow sequence. Bottlenecks resulting from peak workload, staffing, and other resource limitations are identified. Overall operations performance improvements resulting from new processes and information services are estimated in selected processes for specific technical changes.

Figure 11.4 illustrates the application of the workbench to operations for law enforcement "apprehension events" within a given operational scenario, using a

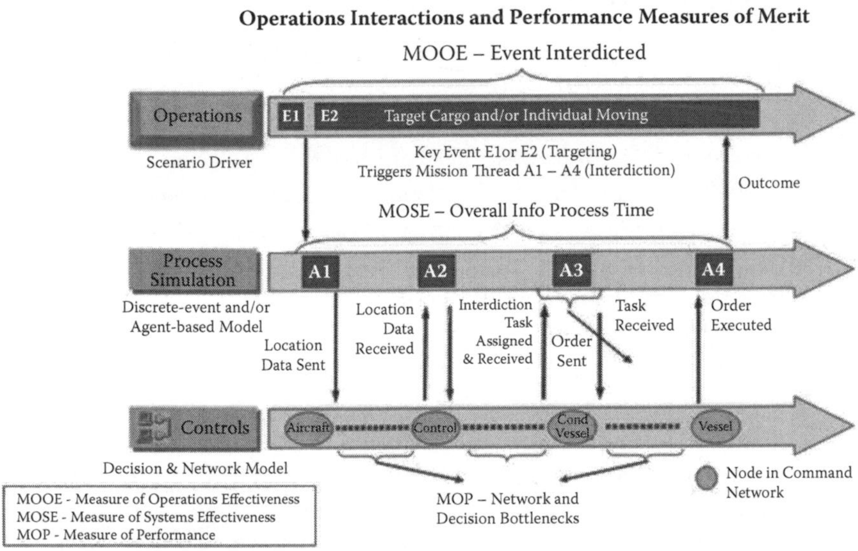

Figure 11.4 Portrayal of performance analysis using the workbench to link operations, response processes, and management control functions for a law enforcement event (i.e., illegal entry at a coastal port represented as "apprehension events" in Figure 11.1). (Reprinted from K.C. Hoffman, T. Pawlowski, D. Payne, and K. Zheng, in International Enterprise Distributed Object Computing Conference (EDOC) Workshop, IEEE Computer Society. © IEEE 2005. With permission.)

process simulation model to describe the multiagency reactions and a network control and decision process layer to describe the time delays in both the communications network and the decision processes that control organizational response. In this illustration the performance of the decision-making processes and technical infrastructure indicate that the individual could be identified and apprehended within the given window of opportunity. This example illustrates the capability of the ESE workbench architect-engineering approach to address major organizational and system aspects of performance management within the governance and decision model employed. The organizational interactions required for mission performance may be planned and analyzed using the readiness model.

Figure 11.5 portrays resource and performance assessments for the operations in a dashboard format. The major features of this screenshot of the dashboard are highlighted. The objective is to network participating agencies and commercial entities for effective organizational coordination as well as for sharing critical information. In this example, all agencies met or exceeded the multiagency collaboration objectives and are rated Green.

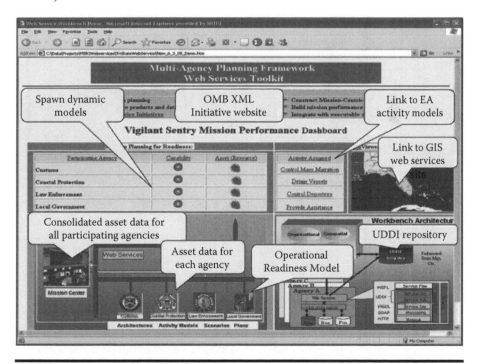

Figure 11.5   Resource and performance dashboard: a web services toolkit for multiagency mission operations planning and analysis. (Reprinted from K.C. Hoffman, T. Pawlowski, D. Payne, and K. Zheng, in International Enterprise Distributed Object Computing Conference (EDOC) Workshop, IEEE Computer Society. © IEEE 2005. With permission.)

11.8.2 Mission/Business Level

Major mission activities in the end-to-end activity sequence are covered in the integrated ESE workbench environment. This integrated approach serves to plan and define the following:

- A complete representation of linked government agency and commercial activities in or related to multiagency missions (e.g., commercial supply chain practices, import and export applications, visa applications for vessel or air crews, port inspection and release processes)
- Organizational roles, responsibilities, authorities, and accountability for specific activities
- Resources (staffing, technology, and information) needed to perform the mission activities at each step for specific workloads and potential areas for reallocation or process improvement
- Gaps or disconnects among activities and organizational responsibilities related to the mission
- Services, databases, and computing/communications infrastructure supporting the activities along with format, protocols, and security/privacy measures for sharing and integrating data
- Structured architectural information for a large portion of the data and information required for performance modeling and analysis.

11.8.3 Services and Components

Business services (e.g., case management, account management, financial management, e-government) and IT services (e.g., commercial web services, data and applications sharing, communications, seat management) are defined and related to specific mission activities. OLAs and SLAs define the specific requirements for information services to support mission performance, whether assembled using commercial-off-the-shelf or custom software components, or as commercially available managed services. This integrated approach:

- Identifies government agency and commercial managed services related to commerce, passenger, and financial accounts
- Provides a basis for integrating similar services and identifying gaps and disconnects that hamper integration
- Provides a basis for comparing and selecting the most effective managed services in developing acquisition strategies, including outsourcing

11.8.4 Data Management and Supporting Infrastructure

Specific solutions and components defined include technologies and systems such as ERP and customer relations management (CRM) systems, data repositories and warehouses, wireless communications, and biometric data collection to implement specific business and IT services. This integrated approach:

- Documents data management and communication services performed by government agencies and commercial partners.
- Establishes a basis for the data-sharing strategy with defined authorities and accountability for data management, along with constraints related to privacy and security.
- Characterizes the computing workload and communications bandwidth required to support specific mission activities; the highest loadings are identified for the retrieval and synthesis of biometric data and for the search of related cargo and financial accounts.
- Guides technology assessments and modernization strategies.

11.8.5 Tool Integration

Several classes of visualization, information management, and simulation tools are applied in the case study. Data sharing and interoperability are enabled by the web services enterprise middleware.

- Operations, processes, systems, and network simulations are federated to provide a comprehensive integration of the environment, enterprise business rules, and enterprise information infrastructure. The federated simulations impose the spatial and temporal dependencies and constraints of the real world, enabling enterprise managers to perform resource allocation and positioning, manage time–space relationships among entities, and determine the overall timing of decision and directive processes.
- The GIS interface provides planners and models with improved and more accurate geospatial locations. GISs can also provide a wealth of environmental and demographic information that affects the planning exercise and enterprise architecture. GIS technology is further leveraged in the ESE workbench using two MITRE-developed technologies, SimServer (Flournoy et al. 2005) and simulation over GIS, SOGIS (Payne et al. 2003), which provide a convenient and near universally accessible method for providing simulation results to a widely dispersed exercise audience. This is useful for bringing in state and local agency planners and for reaching stakeholders of small- and medium-size businesses.
- A metadata server contains metadata, which are described using RDF and RDF schema. An open-source Java framework, Sesame (openRDF.org 2005), is used for storing, querying, and reasoning with the metadata and for translating agency and simulation federate-specific syntax into syntax agreed on by all agencies. This clarifies entity and action references and helps fuse like observations reported by different exercise participants. A similar approach

has been applied by the U.S. General Services Administration (GSA), which is leading the work on ontology and XML-based web services mark-ups for enterprise architectures. The GSA work will facilitate the search and retrieval of information in enterprise architectures for incorporation into the operations-centric architecture and business process simulation models.

11.9 Conclusion

The ESE workbench includes and integrates the basic tools to describe mission operations, organizational roles and responsibilities, business processes, enterprise resources, and information services and technologies needed to improve capabilities and to support the governance of complex enterprises such as international commerce. The workbench is scalable and extensible, using web services enterprise middleware for information sharing and access to other enterprise engineering and planning tools (e.g., process optimization, resource allocation, decision support) in the integration environment. The basic tools have been demonstrated for governance of the most complex operations, those in the multiagency environment. They are equally applicable to individual agencies and commercial operations with complex operations and processes.

The transformation programs of individual firms and agencies represented in their respective enterprise architectures can be brought together in the operations-centric architecture to evaluate their effectiveness in the overall multiagency mission. The definition and use of a consistent set of activities and services in the architectures and in the simulation performance models supports management concepts of activity-based management (ABM), (Player and Keys 1999), and can also be captured in an ERP system for budgeting, monitoring, and management.

Perhaps more important than the feasibility and value of the ESE workbench is the concept of an independent program planning and analysis office for a multinational and multiagency enterprise such as international commerce. The independent "notional" joint office postulated in this case study is a feasible concept that would provide advisory value to all stakeholders.

References

Bunting, W. 2005 (April). Hierarchical performance model, line of sight from modernization tasks to business outcomes and performance metrics. MITRE Technology Program Report.

CIO (Chief Information Officer) Council. 2001. A practical guide to federal enterprise architecture (February).

DoD (U.S. Department of Defense). 2003. *DoD Architecture Framework*. Vols. I and II (15 August).

Flournoy, R.D., Mikula, R.V., Weatherly, R.M., and Seidel, D.W. 2005. SimServer: Simulated data streams on demand via the Web. Presented at *Simulation Interoperability Workshop at MITRE*, McLean, VA, April.

Hamaker, S. 2003. Spotlight on governance. *Information Systems Control Journal*, 1. http://www.isaca.org/Journal/Past-Issues/2003/Volume-1/Documents/jpdf031-SpotlightonGovernance.pdf

Hoffman, K.C., Pawlowski, T., Payne, D., and Zheng, K. 2005. Enterprise business, computing, and information services in a multi-agency environment: A case study in enterprise architect-engineering. In International Enterprise Distributed Object Computing Conference (EDOC) Workshop, IEEE Computer Society.

Milly, P.C.D. et al. 2008. Stationarity is dead: Whither water management? *Science,* 319: 573.

OMB FEA (Office of Management and Budget Federal Enterprise Architecture). 2003. Reference Models Series for Enterprise Architectures (Draft), Performance, Business, Systems and Components, Data and Information, Technology (1 February).

openRDF.org. 2005. *Home of Sesame* (14 July). http://www.openrdf.org.

Payne, D., Hoffman, K.C., and Zheng, K. 2003. Using interoperable process models in a multi-agency planning toolkit for enterprise and C4ISR architecture analysis. Presented at the Simulation Interoperability Workshop at MITRE, McLean, VA (September).

Player, S. and Keys, D.E. 1999. *Activity-Based Management.* 2nd ed. Hoboken, NJ: John Wiley & Sons.

Sowell, P.K. 2005. A readiness model for multi-agency interaction. MITRE Technical Report, McLean, VA.

Zachman, J.A. 1987. A framework for information systems architecture. *IBM Systems Journal*, 26(3): 276–292.

Zheng, K. and Hoffman, K.C. 2006. A service-oriented architecture in a multi-agency environment: A case study in enterprise dynamics. In International Enterprise Distributed Object Computing Conference (EDOC) Workshop, IEEE Computer Society.

Chapter 12

Energy and Materials Systems as an Enterprise Systems Engineering Application: Planning and Analysis for the Economy's Infrastructure

Bradley C. Schoener, Samuel G. Steckley, David H. Reid, Patrick B. Mahoney, Daniel B. Chamberlain, and Kenneth C. Hoffman

Contents

12.1 Introduction

This application provides an example of an enterprise systems engineering (ESE) challenge at the largest scale, crossing major materials and energy-related sectors of a nation's economy. The objective is to illustrate a comprehensive ESE framework for planning, analysis, and management of the use of natural resources for materials and energy by private and public sector stakeholders at local, national, and international levels. Energy and material resource extraction, processing, and end-use activities are private sector responsibilities and operate under regulatory policies. Taking the entire system as the enterprise provides a perspective on the physical aspects of economic activity, the dual space in mathematical programming terms to the monetary economic perspective of national accounts and gross domestic product (GDP). The objective of the framework is to illuminate the complexities of resource use and emerging technologies and their social, economic, and environmental impacts. It provides an information base for industry and government managers for purposes of informing stakeholder investment decisions and policies using decision methods appropriate to the complexities and uncertainties they face.

This natural resource systems study is the largest and most comprehensive example of an enterprise covered in this sourcebook. It encompasses major elements of a functioning economy and the planning and management of government and private policies and investments by stakeholders. The need to understand the local, national, and, indeed, global implications of their collective actions represents a very complex ESE challenge.

The vitality of the economy of a nation or region is based on the effective use of material resources such as cement, iron, aluminum, plastics, wood, and agricultural products for public and private infrastructure, including housing, roads, railways, pipelines, utilities, agriculture, medical facilities, and industrial plants. The associated requirements for labor and energy resources, as well as the environmental impacts, are related to the specific material resources used and the technologies employed to convert material resources to useful products and services in a developed or developing economy. A value chain approach based on the reference material information system (RMIS), described below as a state-of-the-art information system, can be used to provide an integrated framework for information on material resources and finished materials markets to support infrastructure planning and the efficient allocation of resources.

The use of state-of-the-art information technology involving relational databases, object-oriented databases, geographic information systems (GIS), and simulation or optimization models makes information more accessible and useful for planning and analysis. This chapter outlines an approach to a comprehensive integrated materials information system for infrastructure planning. The objectives of this comprehensive approach are to support integrated planning across the entire materials system and to identify and analyze risks and uncertainties that must be resolved during the planning and implementation of materials and infrastructure projects. The scope and content of such a framework is reviewed with examples from the materials and energy sectors. The integration of these sectors into the overall economy is also discussed.

From the perspective of private profit-seeking firms, as well as government policy makers and regulators, there is a need for readily available and high-quality information on material requirements, small and large-scale technologies, markets, prices, and other physical resources employed to improve the level and quality of the infrastructure. Such information is critical to the efficiency of markets, and is particularly important to developing economies and to those in transition to market economies.

Much attention is paid to gathering and interpreting financial statistics on economic development; however, little attention is paid to developing appropriate information on the physical flow of material resources into the infrastructure and capital formation in regional or national markets. There is a need for open sources of technical information accessible to modelers and decision makers that is open to review and critique by advocates and those with different views or data. The RMIS, and associated data on the physical infrastructure, may also be used as a framework representing the annual flow and the stock level in the capital portion of national accounts.

This chapter describes a case study (Hoffman 1995) that applies an integrated information system encompassing critical data elements. It provides a comprehensive base for a variety of both process and econometric modeling approaches to support business decisions and government policy making. The potential roles of process simulation, optimization, and econometric modeling are described.

12.2 Description of the Enterprise—The Materials and Energy System[*]

The material system of a nation encompasses the supply, conversion, and utilization of physical materials such as cement, iron, aluminum, wood, and plastics used

[*] Hoffman, K.C. 1995. An Integrated Materials Information System for Infrastructure Planning. © 1995 Kluwer Academic Publishers: *Journal of Systems Integration*, 5(2): 91–105. Reprinted with kind permission from Springer Science+Business Media.

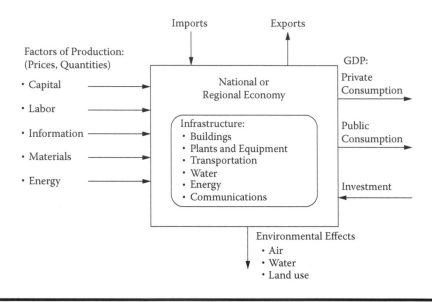

Figure 12.1 High-level representation of the physical aspects of economic activity including the factors of production for economic transition and development. (Reprinted from K.C. Hoffman, *Journal of Systems Integration* 5(2): 91–105, 1995. With permission.)

in constructing and maintaining the physical infrastructure. The energy system is sometimes classified within the materials sector; however, energy system analysis is rather mature from the perspective of databases and analytical models. The notion of a "materials system," on the other hand, has not received nearly the same level of attention with related subsystem research emerging in the 1980s and 1990s (Fischer-Kowalski 1998; Brunner and Rechberger 2004). In view of the relationship between the materials system and the building and maintenance of the physical infrastructure of a nation or region, it is timely to address infrastructure needs and material resources in a comprehensive manner. Principal elements of the physical infrastructure addressed here include housing, roads, railways, pipelines, healthcare facilities, industrial plants, power plants, and refineries. A high-level view of the role of the materials system in the context of the economy is shown in Figure 12.1.

The effective construction, operation, and maintenance of the physical infrastructure requires attention to raw material requirements, processing capacities, and delivery in suitable forms for construction. Logistics issues are also important in timing and scheduling appropriate provisions in the right place, in the right form, and of the right quality. Logistics planning and analysis can draw on the type of comprehensive information discussed here, but have several specialized requirements that go beyond the scope outlined here. The emphasis in this chapter is on providing information for the planning of infrastructure projects and related material recovery, processing, and manufacturing processes to support economic transition and development.

Capital and information are essential inputs to economic transition and development. These are critical factors in development and have a strong influence on the successful planning, implementation, and operations of the economy as well as the energy and information infrastructure needed to provide them. Financial and information activities must also be carefully planned and implemented; an information architecture for financial processes was described by Hoffman (1993).

Increased attention is being given to the definition of a capital account for a region or nation. A capital account would track the public and private investment in capital assets on an annual basis, and would document the total installed value of those fixed assets including the physical infrastructure.

Although the information and communications sectors are critical to economic growth and may use some critical materials, they are not materials intensive in their construction and operation. It should be noted that despite the small volumes of materials involved, a fire in an epoxy plant (Pollack 1993) that shut down production of the grade of this material used in computer chips had a significant market impact. Nevertheless, the emphasis here is the planning and implementation of material resource-intensive infrastructure that requires a reliable supply of energy and material resources.

The data integration and modeling approaches described in this chapter follow a path that has been successfully implemented in the energy sector (Hoffman, Doernberg, and Hermelee 1979; Kavanagh 1979). Major energy system analysis activities were undertaken at Brookhaven National Laboratory, The MITRE Corporation, Stanford University, the Department of Energy, the International Institute for Applied Systems Analysis, and the European Community. These and other activities are summarized in the proceedings edited by Kavanagh (1979). The sectorwide approach to energy planning and analysis, including the integration of energy process models with economic models, was given high priority around the world in the 1970s in response to the "energy crisis" and the concentration of responsibility in energy agencies.

The book by Lee et al. (1990) summarizes the lessons learned from the "energy crisis" and outlines a systems approach to integrated energy systems. There are no similar responsibilities for materials planning and policy and there is no perceived materials "crisis". Nevertheless, there are many examples of infrastructure projects that failed to meet their objectives because they did not fit into the economic system, or were not sustainable due to energy or materials constraints. Materials, including energy, are also critical to the successful planning, implementation, and operation of a nation's infrastructure and can benefit significantly from integrated planning and analysis. Increased attention is being given to the strategic materials and resources for critical technologies and infrastructure underlying the U.S. economy (DoE 2010; DLA 2011; Rogich et al. 2008). The importance of secure supplies of critical resources is particularly important given increased reliance on overseas sources and vulnerable supply chains.

12.3 Statement of the Enterprise Transformation Challenge

The transformation envisioned for this case study stems from the need to produce the physical goods required by a growing economy using natural resources in a more sustainable manner while addressing environmental concerns such as water use and potential climate change.

The Energy Independence and Security Act of 2007 outlined a series of transformation objectives for the U.S. energy system to increase the security of the United States. Elements of this Act included:

- Increasing the supply of alternative fuel sources by setting a mandatory Renewable Fuel Standard (RFS) requiring fuel producers to use at least 36 billion gallons of biofuel in 2022.
- Reducing U.S. demand for oil by setting a national fuel economy standard of 35 miles per gallon by 2020, which will increase fuel economy standards by 40% and save billions of gallons of fuel.
- Requiring that all general purpose lighting in federal buildings use Energy Star® products or products designated under the Energy Department's Federal Energy Management Program (FEMP) by the end of Fiscal Year 2013.
- Updating the Energy Policy and Conservation Act to set new appliance efficiency standards that will save Americans money and energy.
- Establishing an Office of High-Performance Green Buildings (OHPGB) in the U.S. General Services Administration.

The Energy Information Administration (EIA 2007) has a state-of-the art assemblage of information and energy models to gauge the impacts on energy supply–demand balances. The Act will also have major effects on the transformation of other sectors of the economy to produce biofuels from agricultural products, composite materials for lighter-weight vehicles, batteries for hybrid vehicles, and "green" building materials. Analysis of these transformations requires a broad perspective on materials and energy systems and their interactions. In turn, agricultural impacts, for example, will raise water, land use, and significant trade issues.

There are additional demands on the economy that are very materials intensive, such as replacement of ageing infrastructure, including bridges, pipelines, and water supplies. The European community is emphasizing energy and communications infrastructure as an investment priority for the future. Infrastructure issues are also a significant factor for nations in transition to a market economy and in most developing countries. When viewed from a global perspective, these collective activities are heavily influenced by regulatory policies that can have a major impact on material markets and competition among firms and nations. Such impacts must be analyzed and understood in policy formulation, program planning, and investment.

Beyond these examples, the complexities of the materials and energy system include the large and distributed scale of operations; multiagent roles and responsibilities; and associated governance challenges for planning and investment; technical and system diversity of technologies, policies, and management methods; emergent behaviors; security concerns; and major uncertainties raised by ever-changing energy sources and markets. Milly et al. (2008) states that "stationarity is dead." His statement refers to water management as usage increases to levels where availability is affected by the always present vagaries of weather and competing demands, but applies in an even stronger way to the very wide range of factors to be addressed by ESE. Many systems engineering efforts assume a level of "stationarity" to simplify the problem, an assumption that cannot generally be made for ESE challenges. Stationarity assumptions are found particularly often in energy and materials sectors given the high capital intensity of the technologies and systems and the associated expectation of longevity of the invested technologies.

An ESE perspective is required in this very complex enterprise space, and at multiple organizational levels, to characterize and analyze the impacts of decisions and technology choices that lead to sustainable economic and social development.

The expansion in the scope of resources and industrial sectors involved in such a major transformation of the U.S economy requires an improved information base for planning and decisions by all stakeholders. The specific objective of this case study is descriptive to illuminate and formulate a comprehensive information system—the reference materials information system—and demonstrate its potential use by multiple stakeholders to project and evaluate alternative policies, technical strategies and solutions, and socioeconomic–environmental impacts. These capabilities will support entities and programs engaged in the development, acquisition, and implementation of technologies and materials/energy processing systems.

The comprehensive description of the system and its complexities emphasized in this application is an important foundation for planning and decision actions of the multiple stakeholders. Decision methods and theories described in other case studies can be drawn upon as necessary and fitting, considering their specific roles and responsibilities.

12.4 Characterization of the Natural Resources and Materials System

The RMIS provides a comprehensive activity and process description, or value chain, for the flow of physical materials from the resource extraction step through all of the refinement and conversion processes required to produce a material form

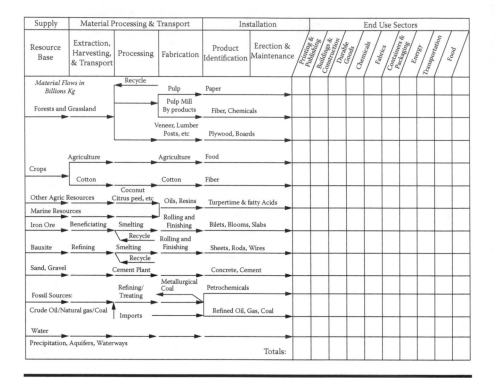

Figure 12.2 The reference materials information system, an activity and process flow template description of the conversion of renewable and nonrenewable resources to finished materials for the physical infrastructure. (Reprinted from K.C. Hoffman, *Journal of Systems Integration* 5(2): 91–105, 1995. With permission.)

that can be used to create infrastructure. This approach was applied in a national study of the potential role of renewable resources to substitute for materials uses of petroleum products (NRC 1976).

The scope and content of the RMIS shown in Figure 12.2 for the U.S. materials supply chain is end-use based and was originally developed by Bhagat and Hoffman (1980) for the NRC study. This format can display a regional or national supply–demand balance. In planning and analysis applications, it is populated with data for a base year and projected to target years, using a variety of systems engineering methods and models discussed below.

The definition of the materials system at this level of detail provides a comprehensive framework for the integration of data at all stages of the value chain. These parameters provide key elements of the state-space of the enterprise:

- Resource data on the location and quality of renewable and nonrenewable sources, and rates of extraction
- Technology data on existing and proposed technologies for extraction, conversion, and processing of materials, and levels of introduction over time

- Cost and price information and projections for investment analysis
- Quantities of labor, energy, and other resources required at each step in the material chain
- Specific utilization and substitution patterns among finished materials and their attributes for use in the appropriate elements of the infrastructure
- Level of demand for energy services and mix of energy forms employed
- Policy incentives and disincentives such as regulations, grants, subsidies, and taxes
- Specific environmental factors and levels at each step in the processes, for example, land use, solid waste, and emissions to air and water, and specification of the control technology and its cost

Each activity link in the RMIS can be characterized by a flow parameter, a processing or conversion efficiency, and a set of coefficients for cost, labor, and energy requirements, as well as air and water emissions, and land use per unit of output flow. The geographic locations must also be specified in this overall regional or national balance. Other critical or valuable materials such as cobalt and chromium for alloying, platinum or palladium catalysts, and so on may also be represented as coefficients related to the appropriate processing activity.

Control parameters that affect the evolution of the enterprise state include:

- Levels of research and development in technologies
- Levels of investment in implementing technologies and systems
- Organizational change
- Policy changes that affect use patterns and investments

When completed for a base year using historical data, projections may be made consistent with economic growth trends and planning objectives to indicate:

- Infrastructure development patterns
- Resource requirements
- Import and export patterns for resources and finished materials
- Resource and material processing bottlenecks
- Recycling opportunities
- Land, water use, and environmental emissions
- Capital and operating costs

The specific enterprise state-space and control parameters applied in a given study are dependent on study objectives. Parameters will be selected and defined with greater precision and, when required, regional dependencies to address issues and questions of direct interest. Whatever the selection, the more comprehensive enterprise parameters are implicit and must be considered in defining assumptions. This makes a strong argument that a very comprehensive characterization of the enterprise, a comprehensive information base, be developed with greater emphasis on physical characteristics such as in the RMIS.

12.5 Strategies and Technologies Examined for the Transformation

The individual firms and agencies engaged in the supply, processing, and use of natural resources make decisions and investments in response to demands and regulatory policies that drive the transformation of the materials and energy system. This transformation must be directed at improved performance levels for industry and the national economy, and must address national and global policies that balance economic and social development with sustainable resource use and environmental challenges. Decisions and actions are sometimes coordinated through government initiatives but often are taken independently for local reasons, adding to the complexity.

Many alternative technology and strategy options are available for transformation of the materials and energy systems. Regardless of the options selected, achieving rational policies requires a robust governance structure operating on a comprehensive information base. An architecture-based approach to governance can help resolve complexity and the many organizational accountability and control issues that arise. This approach ensures that decision processes are integrated across multiple agencies and are responsive to strategic direction and assigned responsibilities. Specifically, the level of information sharing and provisions for accountability, information quality, privacy, and security are tied closely to the decision processes and regulatory policies.

12.6 ESE Analytical Approach

Analyses of the physical aspects of the materials and energy system can be supported using a GIS display of data including plant sites, material transport flows, and emission maps. The International Institute for Applied Systems Analysis has developed and applied a comprehensive set of regional environmental models using these data presentation and modeling approaches (Brouwer et al. 1991).

An example of a more detailed RMIS process description for a specific end use is shown in Figure 12.3 (Schoener et al. 2008). The end-to-end physical flow of materials from agricultural feedstock to ethanol as a vehicle fuel is shown. Inputs of land, fertilizer, energy, and capital are described along with outputs of fuel, by-products, and effluents to air, land, and water.

12.6.1 Material System Process Models

The process structure of the RMIS maps directly into discrete event simulation models and large-scale optimization models. The strengths of process models include:

- The ability to represent material flows, labor, and other factors in physical terms
- The characterization of specific technologies for extraction, conversion, and processing
- Conformance to the laws of science and engineering

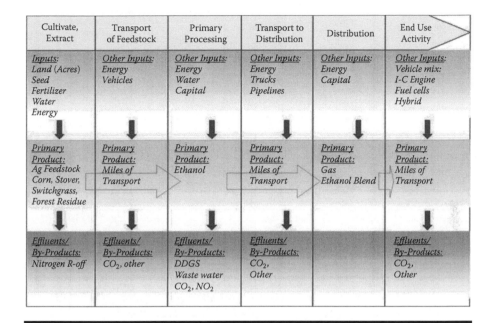

Cultivate, Extract	Transport of Feedstock	Primary Processing	Transport to Distribution	Distribution	End Use Activity
Inputs: Land (Acres) Seed Fertilizer Water Energy	*Other Inputs:* Energy Vehicles	*Other Inputs:* Energy Water Capital	*Other Inputs:* Energy Trucks Pipelines	*Other Inputs:* Energy Capital	*Other Inputs:* Vehicle mix: I-C Engine Fuel cells Hybrid
Primary Product: Ag Feedstock Corn, Stover, Switchgrass, Forest Residue	*Primary Product:* Miles of Transport	*Primary Product:* Ethanol	*Primary Product:* Miles of Transport	*Primary Product:* Gas Ethanol Blend	*Primary Product:* Miles of Transport
Effluents/ By-Products: Nitrogen R-off	*Effluents/ By-Products:* CO₂, other	*Effluents/ By-Products:* DDGS Waste water CO₂, NO₂	*Effluents/ By-Products:* CO₂, Other		*Effluents/ By-Products:* CO₂, Other

Figure 12.3 **Physical input–output flow model for production of ethanol from agricultural feedstock.**

Process tools may be used as appropriate for specific planning and analysis applications. Their principal limitation is the lack of a basis for estimating market demand relationships, thus the need for supplemental economic models.

12.6.2 Relation to Economic Models[*]

There is a direct and natural link between a physical process flow model and macroeconomic planning models as indicated in Figure 12.4. A macroeconomic model with reasonable sector detail (e.g., more than 10 sectors) can be formulated to address physical demands on the materials and energy infrastructure as a function of demographic, financial, and policy factors.

A variety of methods has been employed for the economic analysis of materials and energy resources including econometric methods, linear programming, input–output, dynamic simulation, and systems dynamics. Each method brings some unique perspective to the problem and relative advantages and disadvantages in specific planning and analysis issues.

Econometric methods have strength in the scope and completeness of economic relationships that they capture, but are limited by the form and availability of historical

[*] Hoffman, K.C. 1995. An Integrated Materials Information System for Infrastructure Planning. © 1995 Kluwer Academic Publishers: *Journal of Systems Integration*, 5(2): 91–105. Reprinted with kind permission from Springer Science+Business Media.

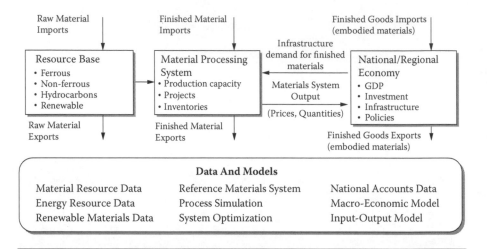

Figure 12.4 **Scope of the materials system, including the resource base, materials processing, and end use infrastructure and its relation to the physical infrastructure underlying national and regional economic development. (Reprinted from K.C. Hoffman, *Journal of Systems Integration* 5(2): 91–105, 1995. With permission.)**

data. Supply and demand models may also be limited by the historical range of experience with price–quantity relationships. When price–quantity relationships change abruptly, either due to external events, shocks, or specific actions, the econometric model may no longer be valid. Linear programming process methods are strong in capturing a large amount of technical and financial data in a framework that has both a physical and economic basis (in the primal and dual forms of the algorithm). It may, however, be limited by the scope of economic and behavioral relationships that may be represented; this deficiency can be overcome by combining the econometric and linear programming methodologies. Such an approach was demonstrated in models of energy–economic relationships (Hoffman and Jorgenson 1977), where the introduction of new technologies and changes in energy resource prices were treated in the process model and macroeconomic impacts were estimated in the econometric model.

The combined linear programming–econometric modeling system provides the capability to represent specific structural and technological change in materials and resource use in an explicit manner, while retaining the overall economic and behavioral relationships captured by econometric methods.

Dynamic simulation and system dynamics methods provide complementary tools for materials analysis; however, there have been examples of misuse of such models due to data limitations and analytical weaknesses in representing valid resource–economy–environment interactions. Nevertheless, given good data at the level of detail represented in the RMIS, and with relationships that are scientifically valid, these methods can be useful. There are numerous examples of applications of these methods in regional energy and resource planning (Zhen 1991; Assimakopoulos 1992).

Interindustry Transactions in the Econometric Model

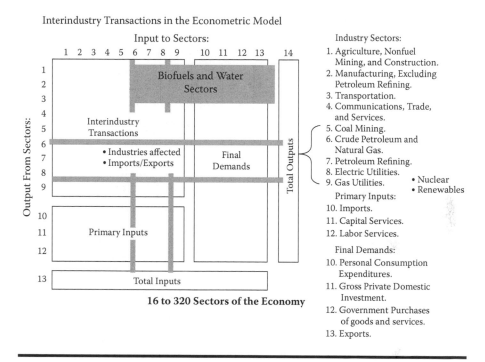

Figure 12.5 presents an input–output model. Text within the figure includes:

Input to Sectors: 1 2 3 4 5 6 7 8 9 10 11 12 13 14

Biofuels and Water Sectors

Interindustry Transactions

• Industries affected
• Imports/Exports

Final Demands

Output From Sectors: 1 2 3 4 5 6 7 8 9 10 11 12 13

Total Outputs

Primary Inputs

Total Inputs

16 to 320 Sectors of the Economy

Industry Sectors:
1. Agriculture, Nonfuel Mining, and Construction.
2. Manufacturing, Excluding Petroleum Refining.
3. Transportation.
4. Communications, Trade, and Services.
5. Coal Mining.
6. Crude Petroleum and Natural Gas.
7. Petroleum Refining.
8. Electric Utilities.
9. Gas Utilities.   • Nuclear
                        • Renewables
Primary Inputs:
10. Imports.
11. Capital Services.
12. Labor Services.

Final Demands:
10. Personal Consumption Expenditures.
11. Gross Private Domestic Investment.
12. Government Purchases of goods and services.
13. Exports.

Figure 12.5   Input–output model of the economy with rows and columns added for biofuels and water sectors.

It is clear that there is no single model or methodology that can deal with all aspects of resource planning and analysis issues of even modest complexity. What is clear is that modeling has often run beyond the data that support the model; hence, the emphasis in this chapter on an integrated materials information system (IMIS) representing both physical and financial data at a detailed process level.

Potential linkages of the RMIS to economic models include an interindustry input–output (I–O) model and a macroeconomic model as illustrated in Figure 12.4. The conceptual addition of biofuels and water sectors to the I–O model is shown in Figure 12.5.

12.6.3 Architecture for the Integrated RMIS*

The RMIS provides a comprehensive framework for resource use, processing capacities and flows, and end use of materials to build and operate the physical infrastructure. This framework also incorporates cost, labor, and environmental coefficients. Current statistics on material supply, utilization, and inventories at var-

* Hoffman, K.C. 1995. An Integrated Materials Information System for Infrastructure Planning. © 1995 Kluwer Academic Publishers: *Journal of Systems Integration*, 5(2): 91–105. Reprinted with kind permission from Springer Science+Business Media.

ious stages in the supply chain should be maintained in conjunction with national accounts for all sectors comprising GDP statistics, a physical national account. Analysts in government, the private sector, and at universities may then develop longer-term scenarios of material flows for policy making, market planning, plant siting, or other purposes.

Physical account data can be maintained in a modern relational database for universal access and ad hoc report generation. Relational databases may be distributed so that responsibility for maintaining the data can be assigned to private sector firms, utilities, and public sector agencies as appropriate.

A full capability GIS can be used effectively to access and display data and analytical results. Most commercial GISs have excellent links to relational databases. They also make use of object-oriented databases for data concerning major facilities or other managed entities. The most advanced GIS applications of this type have been developed for resource mapping, agriculture, and land use planning by national or local agencies; and for automated mapping/facilities management of utility facilities, pipelines, transmission lines, and other infrastructure. The relationship among the materials database, the GIS, and materials models comprising an IMIS is shown in Figure 12.6 along with appropriate linkages for data sharing, modeling, and forecasting.

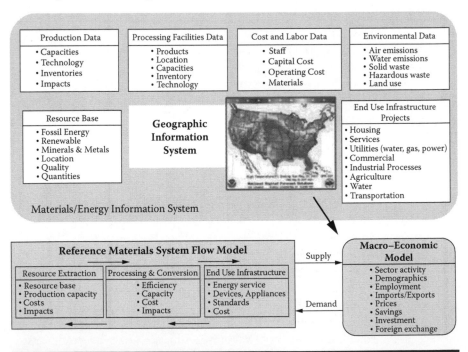

Figure 12.6 The architecture of an integrated reference materials information system. (Reprinted from K.C. Hoffman, *Journal of Systems Integration* **5(2): 91–105, 1995. With permission.)**

12.6.4 Transition to Infrastructure Operations

The data and modeling methods outlined in this chapter are applicable to the planning and implementation of resource and infrastructure development projects supporting economic transition and development. Information systems described are also applicable to the management of resource extraction, materials conversion, and infrastructure operations by responsible utilities and other operators. Extensions to this framework would be needed to deal with logistics, local market, and labor supply issues.

GISs are commonly applied for automated mapping and facilities management applications. Most commercial products incorporate direct linkages to relational and object-oriented databases describing:

- Resources in place and extraction operations
- Facility configuration and capacities
- Infrastructure configuration and operating status (pipelines, transmission lines, communication cables, water systems, etc.)
- Distribution facilities
- Infrastructure development project status

If the methods developed for planning and implementation of materials processing and infrastructure projects follow the recommended approach, they can be supplemented to provide tools for operational management.

12.7 Analysis Results and Recommendations

12.7.1 Applications of the Reference Materials Information System*

The technoeconomic models and methods outlined in the approach have been applied in a variety of studies ranging from material selection for a specific product, to national energy/materials research and development (R&D) planning.

The initial purpose in developing the RMIS was to support the National Research Council Committee on Renewable Resources for Industrial Materials, (CORRIM 1976). The RMIS was used successfully as an integrating framework to document the technical and economic characteristics of concepts proposed by a number of subcommittees. A number of R&D and policy recommendations were made to promote the use of renewable resources as a substitute for petroleum feedstock.

---
* Hoffman, K.C. 1995. An Integrated Materials Information System for Infrastructure Planning. © 1995 Kluwer Academic Publishers: *Journal of Systems Integration*, 5(2): 91–105. Reprinted with kind permission from Springer Science+Business Media.

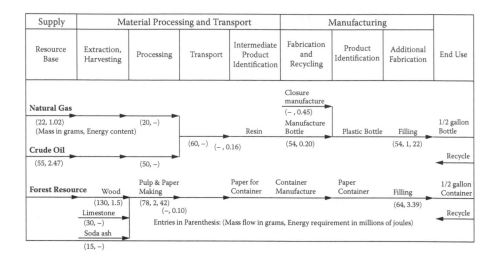

Figure 12.7 RMIS flowchart and data regarding material substitution possibilities for a half-gallon food container. (Reprinted from K.C. Hoffman, *Journal of Systems Integration* 5(2): 91–105, 1995. With permission.)

An early example of material selection using a more detailed RMIS process description for a specific product is shown in Figure 12.7. Here the substitution of renewable resources for petroleum-based plastic materials in a half-gallon food container is analyzed. Given the cost, process efficiencies, and environmental emissions at each step of the alternative processes, the integrated end-to-end impacts of the alternatives may be analyzed. Generally, an analysis such as this provides detailed information; however, decisions and selections must be based on some overall trade-off analysis, in this case between desires to conserve imported oil-based materials versus the local environmental impacts of using indigenous resources.

The physical I–O model for ethanol production shown earlier in Figure 12.3 has been applied in a recent study of the impacts of very large production levels on the energy system, agriculture, consumer, and other elements of the infrastructure. The ethanol RMIS encompasses corn, stover, switch grass, and forest products as sources of cellulose and sugar for ethanol production and is applied in a large-scale production scenario. Both technical and macroeconomic impacts were explored using the socio–techno–economic models described above.

A major outcome of applying the integrated RMIS to analyses of materials and energy supply and utilization is to help identify and analyze risks and uncertainties that must be resolved during the planning and implementation of materials and infrastructure projects.

12.7.2 Materials and Energy Analysis Topics[*]

The specific planning and analysis applications for a materials information system will depend on a number of factors, including political organization of the economy and the stage of development, as well as regional culture and customs. The overall objective of planning and analysis of the physical infrastructure and supporting materials system is to coordinate expansions and improvement projects across all activities in the system to identify and evaluate the capital and operating investment required to reduce or eliminate bottlenecks and disruptions throughout the entire supply, processing, and end-use chain. Investments must be tightly coupled with the scheduled project plans, and the policy environment should work to provide appropriate incentives to industry and engineering firms along the entire chain.

The emphasis here is on coordinated planning of infrastructure expansion projects. Similar objectives may be applied to the operational phase and the system concept provided by the IMIS may also be applied at that stage with the addition of appropriate logistics management methods. The specific policy priorities that are most pertinent to various types of economies are outlined below. Almost every issue listed can arise at any stage of development, but the priorities may vary.

Mature market economy:

- Land use standards and allocation
- Maintenance requirements and investments for the infrastructure
- Definition of critical materials and their form at the appropriate step in the processing chain
- Environmental impacts, local and regional, as well as contribution to global concerns
- Regional labor opportunities and training requirements
- Research and development priorities

Transition to market economy:

- Provision and maintenance of essential infrastructure to support economic transition
- Appropriate sectors for privatization and timing of the transition
- Productive use of indigenous resources
- Pricing of materials during the transition
- Sustainability and resiliency of the materials system and infrastructure to economic and political shocks

---

[*] Hoffman, K.C. 1995. An Integrated Materials Information System for Infrastructure Planning. © 1995 Kluwer Academic Publishers: *Journal of Systems Integration*, 5(2): 91–105. Reprinted with kind permission from Springer Science+Business Media.

Developing economy:

- Productive use of indigenous resources
- Provision and maintenance of essential infrastructure to support economic transition
- Coordination of budget and financial plans with physical construction activities
- Long-term consequences of materials and technologies employed on the development path
- Sustainability and resiliency of the materials system and infrastructure to economic and political shocks

A life-cycle approach must be taken to materials policy. Resource, processing, and utilization policies and regulations should not be made on the basis of only initial cost; maintenance, replacement, and disposal costs must also be taken into account.

12.7.3 Materials and Energy Planning Methodology*

A material resource planning methodology based on the use of the RMIS and the IMIS for the specific case of a developing economy is outlined in Figure 12.8. The features of this methodology (Hoffman and Basile 1982; Hoffman and deTerra 1984) involve three basic elements:

1. Estimates of market demands for materials based on analysis of the overall economy of a region or nation, including specific development projects. This linkage can be developed at an aggregate level, but a disaggregated approach is preferred where planned projected construction activities and development projects are defined.
2. A comprehensive materials information framework as provided by the IMIS.
3. Analytical models including:
   - Economic forecasting models.
   - Construction plans.
   - Resource allocation and optimization models.

The logic of this approach to sequential planning is clear. A key step is that of project definition, where (1) specific projects are identified to remove projected development bottlenecks (e.g., provide increased cement production capacity); (2) energy needs are identified (e.g., build or supplement the capacity of power plants to provide for material processing); and (3) a specific trained labor force

---

* Hoffman, K.C. 1995. An Integrated Materials Information System for Infrastructure Planning. © 1995 Kluwer Academic Publishers: *Journal of Systems Integration*, 5(2): 91–105. Reprinted with kind permission from Springer Science+Business Media.

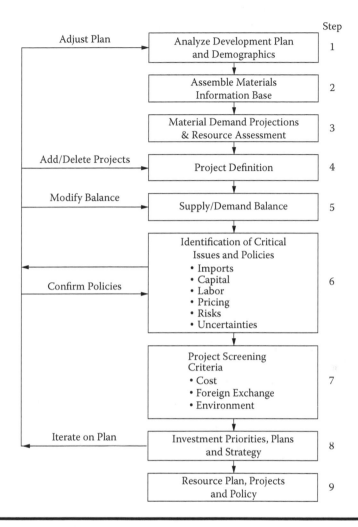

Figure 12.8 Planning methodology addressing planning and analysis for resource development, materials processing, and infrastructure projects. (Reprinted from K.C. Hoffman, *Journal of Systems Integration* 5(2): 91–105, 1995. With permission.)

requirement is identified. The impact of defined projects on the supply-demand balance leads to financial plans and implementation plans. Iteration is required back to the project definition step to add, delete, or redirect projects and to set priorities.

12.7.4 Materials and Energy System and Infrastructure Risks

Risk is inherent in all materials systems and infrastructure programs and is a major factor that must be addressed in the planning and implementation phases, as well as in subsequent operations. Haefele (1991) has published extensively on risks in the energy sector, as well as on the subject of energy systems analysis; there is much in common with the overall materials systems risks. The magnitude of risk is dependent on the technologies employed (stable, emerging, new) and on the stability of the materials systems and markets. The major risk categories that are encountered are outlined below. Idealized conditions are used as examples to illustrate risk avoidance measures but are never realized fully in practice.

Policy and business risk: Public policies generally reflect the public interest and preferences that balance economic development with environment, health, and safety concerns. Policies must also provide a clear and stable regulatory environment as a basis for investment. Private sector responsibilities in the industry–government partnership must be defined and understood by all parties for all materials system activities.

Resource risk: Material resources must be adequate to support materials processing and infrastructure projects through their entire life cycle, and must satisfy all other projected demands on those resources. Financial resources for construction and operation must be available with stable long-term commitments. Trained human resources must be available, or a training plan must be implemented to ensure a trained labor supply.

Project risk: Project objectives must be clearly defined and planning must be integrated to ensure stable supplies of materials, energy, and labor to carry out the project. Design and operating requirements must be stable. The potential for schedule slippage exists on every project. The resulting program risk is much more than the simple sum of project schedule slippages as they are compounded by the cascading effect of interrelationships and interdependencies between related projects in other portions of the materials system. Contractors must have referenced experience in the projects to be undertaken.

Technical risk: Proven processes and technologies must be employed to minimize technical risks. Component obsolescence and degradation can constrain system performance as technical components do not meet expectations or specifications. A maintenance and upgrade plan should be in place to ensure continued operability. The performance of individual projects is important; however, all related projects and activities must perform to meet overall end-to-end system objectives. Information systems used for control of processes always present risks that merit special attention. Computers are available that are fault tolerant, and other backup measures may be taken to ensure the reliability of data and control applications. A reliable communication system must be available with provision for emergency power supplies during power outages. There should be prior experience with any software in similar applications and the software should be fully documented.

Health and safety risk: Environmental impact statements provide the best institutional mechanism to identify health, safety, and environmental risks. These risks must be considered in a local, regional, and in some cases a global context to identify potential impacts and constraining factors.

12.8 Governance

The importance of governance in major programs is stressed by Hamaker (2003), whether managed by government agencies or the private sector. Policy studies and analyses must address specific stakeholder responsibilities at the appropriate level and scope. Well-defined governance roles and responsibilities facilitate support functions through:

- Strong accountability for mission activities performed under local and central direction
- A virtual support structure with central mission integration, but distributed support resources and information

The application of enterprise-level models can support stakeholder responsibilities and accountability. Responsibilities are matched to the hierarchy of private sector and public entities in Table 12.1. The responsibilities of supporting agencies are noted. For the roles and functions indicated, the responsible entities are the principal managers and users; they are also accountable for their currency of planning materials.

The coordination of these governance elements does not come easily. All of the technical elements are the subject of detailed systems engineering analysis and design. Architectures can play an important role in governance as described here, but require specific additions to their scope and content as dictated by the specific roles and objectives. The RMIS supports the required information for coordination.

In the case of the energy system, the Department of Energy's EIA has legislative responsibility to maintain current data and to develop projections of future energy supplies, demands, and prices. Similar information is collected on a global basis by the International Energy Agency in Paris. These information sources and projections are used by all parties. The private sector entities have their own information sources and outlook and use them along with government sources in their business and investment plans.

Responsibilities for materials and natural resources information are spread across multiple agencies (U.S. Geological Survey, Interior, and Commerce); a comprehensive integrated information base is a much needed capability in modern society.

12.9 Conclusion

The RMIS includes and integrates the information to describe physical and economic aspects of the natural resource supply and use, organizational roles and responsibilities, business processes, enterprise resources, and information services and technologies needed to improve capabilities and to support the governance of this complex enterprise. There are major data gaps in the representation of the physical materials activities, processes, and flows. This information is very important in

Table 12.1   Summary of Governance Roles and Responsibilities for Materials and Energy Systems

Functional Hierarchy	Governance Role	Responsible Entity
Natural resource management	Policy formulation	Government
	Regulation	Local and federal
	Resource development	Private sector
Demand sectors	Consumer expenditures	Public
Consumer demand	Product safety	Government
Public works	Public Investments	Government, all levels
Government operations	Government demand	Government, all levels
System development and operations	Extraction	Industry
	Processing and conversion	Utilities and producers
	Delivery (pipelines, grid)	Utilities, Independent System Operators (ISOs), Regional Transmission Organizations (RTOs), Federal Energy Regulatory Commission (FERC)
	Planning	
		Industry and government
Services and components of systems, technology	Materials	Industry
	Energy	
Research	Near term	Industry
	Long term, high risk	National laboratories, government–industry collaboration
Data and information	Resource base	Government, USGS, USDA, DOI
	Systems	Government, DoE
		Industry, utilities
Technology selection	Utility management	Utilities, industry

defining efficiency and environmental parameters and cannot be ignored in policy analysis. It is far preferable to develop even rough estimates of such critical factors and subject them to sensitivity analysis than to ignore them in the interest of constructing "validated" models.

A retrospective validation of early long-term energy projections, circa 1970s, using these methods published by Hoffman (2011) indicates strong descriptive power in capturing the opportunities for energy efficiency that led to lower growth rates in energy demands. The methods also had significant predictive strength highly conditioned, of course, by assumptions that went into the analyses.

The development and maintenance of the physical infrastructure of a nation or region is essential to successful economic transition and development of market economies. Successful market economies rely on a free flow of high-quality information on local resource markets and processing capabilities, and inventories. Much of the current emphasis is on financial market data; physical data on quantities produced, quantities in inventory, and quantities of finished goods are less accessible. A review of statistical abstracts for several years indicates that much of the data on material supplies and finished product required for the RMIS can be derived; however, the processing and inventory parameters must be based on a more thorough analysis and reconciliation. In particular,

- National accounts and statistics should address requirements for physical data on materials for private sector markets and responsible government agencies.
- An integrated planning methodology must be employed to ensure that projects are viable in the context of the overall system and that all potential risks are identified.
- The systems methodology can reveal inconsistencies and imbalances in market data that must be understood by all participants in a project.
- Uniform approaches to materials system physical accounting should be developed and promulgated among trading partners.

Projects and policies often fail due to deficiencies in the transportation, energy, water, and communications infrastructure. A large number of these must be considered to be "system failures" and indicate that the system configuration, capacities, and bottlenecks must be defined and managed by private and public sector institutions. For example,

- The physical infrastructure must be defined, and its configuration and requirements published using a systems approach.
- All critical labor, material, and energy requirements for infrastructure development and operation should be identified and their markets should be tracked.

■ Government policies affecting projects and private sector activity should be based on realistic physical and financial information. Policies and regulations should be based on a cost–benefit analysis taking technological capabilities and costs into account, and reflecting likely market responses.

All pertinent materials system parameters must be analyzed for planned and proposed projects at the definition and evaluation stage, including:

■ Physical material flows
■ Labor requirements
■ Strategic materials required
■ Life-cycle implications
■ Trade implications
■ Environmental impacts of full implementation

Long-term consequences of full implementation of resource allocations and processing technologies should be analyzed. The scope of analysis may also be enlarged beyond the national or regional focus to a global market focus when the projects are very large, are implemented in many areas, or involve critical materials with uncertain supply prospects. One should ensure that:

■ The selected project and large-scale implementation path is sustainable for the long term.
■ The costs, requirements, and impacts of the alternatives are identified and estimated for the entire project life cycle from construction, through its operation, to shutdown and dismantling.
■ Externalities are identified (e.g., environmental, trade, and labor impacts).
■ Access and usability of the IMIS can be promoted by the use of state-of-the-art information and communications technology.
■ Data should be organized in a relational and object-oriented format for accessibility and ability to support ad hoc reports and analyses.
■ A geographic interface should be provided for access and analysis.
■ System simulation or optimization with animation is feasible and should be used to help make the information more useful for policy formulation and decision making.
■ Material system models should be integrated with economic policy models.

The physical material resources of the earth are large as is the potential labor force. There is a wide range of proven processing and construction technologies including small-scale technologies adaptable to local environments. At the same time, the needs of the world's population for comfort, sustenance, and security are almost overwhelming and there is much suffering and tension. The infrastructure and supporting materials supply systems are important elements in meeting these needs. For proper resource allocation, these systems must function within the overall economy of a nation or region that is often constrained by inadequate savings and investment, and by shortages of trained personnel in the marketplace. Accurate, reliable, and timely information is essential to the efficiency of markets, the efficient allocation of resources, and decentralized decision making.

It is clear that close and continuous coordination must be maintained between financial planning and policies addressing markets, and the physical projects planned and implemented by the private sector, international agencies, and aid organizations. A comprehensive materials information system can support such coordination.

References

Assimakopoulos, V. 1992. Combining decision support tools and knowledge based approach for the development of an integrated system for regional energy planning. *Energy Systems and Policy*, 15(5): 245–255.

Bhagat, N.K. and Hoffman, K.C. 1980. *The Reference Material System—Materials Policy Information System for Renewable Resources, A Systematic Approach*. Saltillo, Mexico: Academic Press.

Brouwer, F., Thomas, A.J., and Chadwick, M.J. (Eds.) 1991. *Land Use Changes in Europe: Processes of Change, Environmental Transformations and Future Patterns*. Dordrecht: Kluwer Academic.

Brunner, P.H. and Rechberger, H. 2004. *Practical Handbook of Material Flow Analysis*. Boca Raton, FL: CRC Press.

CORRIM (Committee on Renewable Resources for Industrial Materials), National Research Council. 1976. Report of the Committee on Renewable Resources for Industrial Materials. Washington, DC: National Academy of Sciences.

DLA (Defense Logistics Agency). 2011. *Strategic and Critical Materials Operations*. Report to Congress (January).

DoE (U.S. Department of Energy). 2010. *Critical Materials Strategy* (December).

EIA (Energy Information Administration. 2007. *U.S. Energy Outlook*. U.S. Department of Energy.

Fischer-Kowalski, M. 1998. Society's Metabolism. In G. Redclift and G. Woodgate (Eds.), *International Handbook of Environmental Sociology*. Cheltenham, UK: Edward Elgar.

Haefele, W. 1991. Energy, risk, and environment. *Energy Systems and Policy*, 14(4).

Hamaker, S. 2003. Spotlight on governance. *Information Systems Control Journal*, 1. http://www.isaca.org/Journal/Past-Issues/2003/Volume-1/Documents/jpdf031-SpotlightonGovernance.pdf

Hoffman, K.C. 1993. Management of enterprise-wide systems integration programs. *Journal of Systems Integration*, 3/4 (September): 201–224.

Hoffman, K.C. 1995. An integrated materials information system for infrastructure planning. *Journal of Systems Integration*, 5(2): 91–105.

Hoffman, K.C. 2011. Perspectives on the Validation of Energy System Models for Long-Term Projections. International Energy Agency–ETSAP Workshop, Stanford University, July 9.

Hoffman, K.C. and Basile, P. 1982. Energy resources and environmental policy: Survey of analytical approaches. In *Proceedings of the First U.S.–China Conference on Energy, Resources, and Environment*, Beijing, November.

Hoffman K.C. and deTerra, N. 1984. Energy planning and long term strategies for developing countries. Presented at the Fifth Annual International Meeting of the International Association of Energy Economists, New Delhi, January.

Hoffman, K.C. and Jorgenson. D. 1977. Economic and technological models. *Bell Journal of Economics,* 8(2, Autumn): 444–466.

Hoffman, K.C., Doernberg, A., and Hermelee. A. 1979. Impacts of new energy technologies as measured against reference energy systems. In *Proceedings of Conference on Non-Fossil Fuel and Non-Nuclear Fuel, Energy Strategies*, Honolulu, January. Oxford: Pergammon Press.

Kavanagh, R., Ed. 1979. *Energy Systems Analysis*. Dordrecht: D. Reidel.

Lee, T.H., Ball, B.C., and Tabors, R.D. 1990. *Energy Aftermath*. Cambridge, MA: Harvard Business School Press.

Milly, P.C.D. et al. 2008. Stationarity is dead: Whither water management? *Science,* 319(1 February): 573.

NRC, National Research Council. 1976. *Renewable Resources for Industrial Uses*. Washington, DC: National Academy of Sciences.

Pollack, A. 1993. Japan factory for making chip material destroyed. *NY Times*, July 5.

Rogich, D. et al. 2008. Material flows in the United States: A physical accounting of the U.S. industrial economy. Report by the World Resources Institute (April).

Schoener, B., Hoffman, K.C., Ibe, Z., and Steckler, S. 2008. Physical input-output model of biofuels production. Report. McLean, VA: MITRE.

Zhen, F. 1991. An SD model applicable to the study of rural energy development strategy in Beijing. London: Taylor and Francis: *Energy Systems and Policy,* 14: 213–226.

Chapter 13

Modeling the Nation's Healthcare System as a Dynamic Enterprise*

Fran Dougherty, Kenneth C. Hoffman, Honora R. Huntington, Joseph K. Jun, Dave Klein, Kristin Lee, Bradley C. Schoener, and Mark Walters

Contents

---

* Sections 13.2–13.4 adapted from Dougherty et al. 2012. IEEE International Systems Engineering Conference. © 2012 IEEE.

13.1 Introduction

This chapter presents a multiscale data analytics approach for modeling the U.S. healthcare system as an enterprise. The U.S. healthcare enterprise encompasses public and private entities with significant government involvement in policies and programs that have widespread interactions and effects at all levels of the system; in other words the U.S. healthcare enterprise is a complex adaptive system. As such, integrated holistic data analytical methods are needed to understand the enterprisewide interactions and outcomes of proposed policy initiatives. An enterprise systems engineering (ESE) approach is applied to organize diverse datasets and perform integrative analyses of patient populations, health services, providers, facilities, technology, insurers, pharmaceuticals, research, and the U.S. economy.

Hybrid models are applied to a selected health service and proposed interventions to affect outcomes and costs. Type 2 diabetes management is used as a case study with selected policy and technology interventions to deal with the lifetime complications and consequences of this condition. The hybrid modeling structure includes agent-based methods to portray the movement of patients in the statespace, systems dynamics to capture policy and programmatic influences, discrete event methods to describe sequential service processes, and economic analysis methods to capture costs and impacts on the healthcare enterprise in the context of the overall economy.

At a higher macroscale, the application of an economic analysis model to policies that expand insurance coverage to the currently uninsured population is presented. The methods described address the need for an enterprise-level system-of-systems (SoS) approach to organize the vast and diverse datasets published by various agencies into an open-source data-analytic framework. The framework encompasses the healthcare enterprise to support the comprehensive analysis of the impacts of policies and programs across the many dimensions and scales. Policy and programmatic areas of interest include insurer coverage decisions for new or modified health services, new technologies for diagnostic, therapeutic, and preventive services, and other major influences on the cost and quality of healthcare services.

Rouse (2008) defines the healthcare system as a complex adaptive system and therefore not amenable to hierarchical decomposition. The hybrid modeling methods applied in this chapter encompass the multiple scales of the healthcare system in an integrated and unified approach for analyzing this complex system-of-systems. The emphasis is placed primarily on methods and illustrative applications to healthcare planning, analysis, and evaluation, rather than on specific results. Analytical results are highly dependent on the purpose, assumptions, and context of the policy studies and are meant to inform and improve then-current decision processes.

13.2 Background on Healthcare Transformation[*]

The healthcare system is a large and comprehensive example of a service-oriented enterprise. Health services and systems range across many sectors of the U.S. economy comprising 18% of our gross domestic product (GDP) in 2010 with an annual growth rate of 4%. This raises major challenges to long-term affordability and sustainability that are central to government policies and programs to transform healthcare.

The complexities of the nation's health sector are evident in policy debates that address quality, access, and cost from the perspectives of the public and the government, and can best be described and analyzed using complex systems methods. This chapter applies line of sight performance modeling concepts (Dougherty et al. 2006), along with an ESE process, to the formulation of a comprehensive reference health services system (RHSS) framework and associated data and multiscale modeling methods. The framework is designed to support planning, analysis, and management of government policies and programs to transform the healthcare enterprise to a more effective and sustainable state. The RHSS framework illuminates the complexities of human and physical resources and emerging technologies to deliver health services. It provides a comprehensive data-analytic framework for use by stakeholders and analysts to inform investment decisions and policies using decision methods appropriate to the complexities and uncertainties of this sector.

Although much of the policy emphasis is on bending the cost curve for health services downward, technological innovation is leading to ever more effective healthcare services that deal with hitherto untreatable conditions; these emergent effects pose additional policy and analytical challenges.

The planning and management of government and private policies and investments by stakeholders and the need to understand the local and national implications of their collective actions represent complex ESE challenges. Policy issues and programs for healthcare transformation range from local initiatives and specific

---

[*] Dougherty, F.L. et al. 2012. © 2012 IEEE. Modeling the U.S. healthcare system as an enterprise—multiscale hybrid data analytic methods. In IEEE International Systems Engineering Conference, Vancouver, B.C., March. With permission.

health services to national efforts to improve effectiveness and manage costs. No single model or analytical method can address this range of policies and programs, hence, the emphasis here is on multiscale hybrid methods that use techniques appropriate to the scale and purpose of the transformation initiative.

The use of state-of-the-art information technology involving relational databases, object-oriented databases, geographic information systems, and simulation or optimization models make information more accessible and useful for planning and analysis. This chapter outlines how the RHSS may be applied as an integrated health services information system for policy and planning. The objectives of this comprehensive approach are to support integrated planning across the entire healthcare system and to identify and analyze risks and uncertainties that must be resolved during the planning and implementation of programs and policies. The scope and content of such a framework is reviewed with illustrative examples applied to policy issues. The integration of healthcare sectors into the overall economy is also discussed.

From the perspective of private profit-seeking firms, as well as government policy makers and regulators, there is a need for readily available and high-quality information on the demographics of healthcare, human resources, and small- and large-scale technologies that provide new health services and that may be applied to control costs. Complex market behaviors are subject to complex cost signals, incentives, and disincentives employed to improve the level and quality of preventive and clinical services. Such information is critical to the efficiency of healthcare services.

Much attention is paid to gathering and interpreting financial statistics on health services; however, it is equally important to develop appropriate information on the human and other physical resources and assets required to deliver high-quality care. The RHSS and associated data can be used as an integrating framework representing the interconnectedness and dynamics of this complex system. It also provides a comprehensive base for a variety of both systems engineering and economic modeling approaches. The potential roles of process simulation, optimization, and econometric modeling are described.

The ESE methodology provides a repeatable process to describe and analyze the behavior of complex systems in response to transformational policies and programs. The process encompasses policy, organizational, economic, and technical (POET) dimensions of the enterprise that must be coordinated for effective transformation.

13.3 Description of the Enterprise— U.S. Healthcare System[*]

The enterprise is defined here as a purposeful social, technical, and economic endeavor designed to create value for its stakeholders involving:

---

[*] Dougherty, F.L. et al. 2012. © 2012 IEEE. Modeling the U.S. healthcare system as an enterprise—multiscale hybrid data analytic methods. In IEEE International Systems Engineering Conference, Vancouver, B.C., March. With permission.

- Policies, organizations (people), economics, and technology interacting with each other and their environment in operational processes as a complex system-of-systems to achieve goals
- Organizations (single or multiagent, and possibly virtual) created for the undertaking
- A readiness to embark on bold new ventures to improve performance and manage costs

The dynamics of an enterprise are highly dependent on the mission or business characteristics (the enterprise landscape) and the perspective on the enterprise required to address management issues and inform decision making (operational, scalar, and temporal perspectives).

The healthcare enterprise is clearly a "purposeful endeavor" and can be described at several levels of detail (scales) ranging from the delivery of specific services to an individual, through the facilities and staff that deliver those services, to the national level of the healthcare sectors within the overall economy. In all cases a complete description of the enterprise must include those sectors responsible for the delivery of services as well as the supporting administrative, insurance, and governance functions. The Department of Health and Human Services and other agencies maintain a comprehensive assemblage of information (Department of Health and Human Services 2011a; Dartmouth Institute for Health Policy and Clinical Practice 2010).

The *healthcare enterprise* is defined here to incorporate those sectors of the economy engaged in the delivery of health services as well as those sectors that provide supporting goods and services. This description of the healthcare enterprise includes the following 12 sectors within the U.S. economy.

- Hospitals
- Physicians
- Nursing homes
- Other ambulatory healthcare services
- Home health services
- Social assistance
- Dental, ophthalmic
- Medical technology
- Pharmaceuticals
- Information systems
- Insurance
- Government

The description of the healthcare system at this level of detail provides a comprehensive framework for the integration of data pertaining to all stages of healthcare services: preventive, diagnostic, therapeutic, and long-term management and maintenance.

The above healthcare sectors are identified specifically in dynamic versions of interindustry models of the U.S. economy, such as the INFORUM LIFT model (2007), that covers 97 sectors of the full economy and describes population demographics and household expenditures across all sectors of the economy including health services and insurance, taxes for government health programs, government expenditures, employer-provided plans, and interindustry product and service flows to healthcare providers. Economic interindustry models proved a basis for near and long projections of healthcare costs and services along with the labor, GDP, and personal expenditure/income effects. These analytical methods are discussed in greater detail below.

The national economic perspective on the healthcare system and related policy issues are described for the economy level in the multiscale portrayal, Figure 13.1. This multiscale representation expands the four-level framework proposed in a National Academy of Sciences study (NAS 2005). It organizes the healthcare enterprise into four levels for analysis. Multiple scales affected by healthcare policies and technology range from the microscale (workforce and health services) to the macroscale (health sectors in the context of the U.S. economy and population demographics). These scales describe a nested enterprise with the identification of attributes that provide direct patient services, support patient services, and influence the cost and quality of services. Programs and policies affect all levels of the enterprise with complex interactions among the elements shown, the ESE challenge.

Detailed descriptions of processes for the delivery of health services are required at the lower scales in order to provide program details to be incorporated into

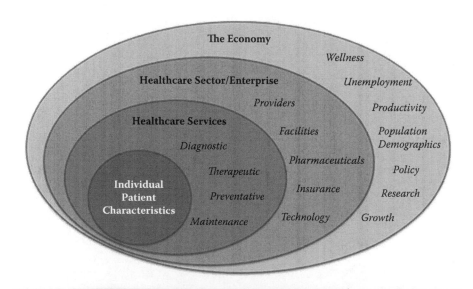

Figure 13.1 Multiscale representation and attributes of the healthcare enterprise. (Reprinted from F.L. Dougherty et al. 2012. © IEEE. With permission.)

economic projections, and to analyze the impacts of specific policies and programs and alternatives. This higher fidelity definition of the healthcare system can incorporate a wide variety of data including:

- Human resources
- Physical resources: hospitals, physician offices, equipment, and supplies
- Information, communication, and sensor systems
- Patient demographics
- Payment streams

13.3.1 Statement of the Healthcare Enterprise Transformation Challenge

The transformation envisioned in comprehensive healthcare policies is to provide greater access to health services for all citizens with more effective outcomes and at a sustainable cost. The Affordable Care Act (ACA 2010) addresses virtually all elements of the healthcare enterprise and has generated considerable analysis and debate. Multiscale analyses can provide insights into the complex interactions of the policy elements that have been and are yet to be defined.

Earlier bills proposed in Congress describe other desired outcomes of transformation of the healthcare sectors. Using a highly focused initiative preceding the major legislation of 2009–2010 as an example, the Health Technology to Enhance Quality Act of 2005 (2005) outlines a series of transformation objectives building on the Electronic Health Record (EHR) and other information technologies.

A report by the Engelberg Center at Brookings, "Bending the Curve" (Brookings 2009), presents ideas for controlling the cost of healthcare, mostly through information technologies. Others have proposed a broad range of technology and process changes as well as major reorganization of the enterprise that challenges healthcare analytics and dictates the application of complex systems methods. Again, attention must also be directed toward emergent behaviors and technological change regarding new medical services that can force the curve upward.

The government role, through direct services provided by the Veterans Administration and Military Health Services and direct financing using payroll fees, account for nearly 10% of the U.S. GDP. The transformation challenge in the Medicare and Medicaid programs is a sustainable balance of costs and revenues, a critical topic in all policy discussions. Medicaid addresses the underserved population directly, with roughly half of the potential beneficiaries actually enrolled. The financial burden of this program on states, in particular, is leading to significant change in services and eligibility.

Approaches to coverage for the uninsured population also include the IRS Health Coverage Tax Credit (HCTC) Program (IRS 2005) administered by the IRS that applies to individuals and families that are receiving benefits from the Pension Benefits Guarantee Corporation (PBGC) or who have lost their jobs and insurance due to foreign competition, and the Massachusetts Legislature program (2006) that requires insurance coverage with reimbursements based on income. Both of these policy approaches are scalable to larger groups of the underserved.

A policy challenge at a national level, and the potential role of informatics, is expressed in Executive Order 13335 (2004): the president directed that the Departments of Veterans Affairs (VA) and Defense (DoD) shall jointly report to the Office of the National Coordinator for Health Information Technology (ONCHIT) on the approaches the departments could take to transform clinical practice and healthcare delivery in rural and medically underserved communities through the use of affordable health information systems. This report recommends the need for a common "blueprint" or "road map" from which all interested parties can proceed. The report further recommends approaches that focus on standards (e.g., data, security, messaging, technical, and communication) and interoperability; infrastructure considerations (e.g., networks, hardware, and software); contracting incentives; technology transfer; and sharing of lessons learned.

Transformation initiatives also emerge at lower scales, such as the New York City (NYC) Department of Health initiative (NYC 2005) to apply electronic laboratory-based reporting of hemoglobin tests in a pilot area in the south Bronx, a poor, largely African-American and Hispanic-American community with particularly high rates of diabetes. Similar programs are planned in Oregon and Washington states to address the $132 billion/year expenditures nationwide for the treatment of diabetes. An ESE perspective is required in this very complex enterprise space to characterize and analyze the impacts of decisions, technology choices, and at multiple organizational levels that lead to sustainable economic and social development.

The transformation of healthcare sectors of the U.S economy requires an improved information base for planning and decisions by all stakeholders. The specific objective of this case study is descriptive to illuminate and formulate a comprehensive information system, the RHSS, and related multiscale models, and demonstrate their potential use by multiple stakeholders to project outcomes and evaluate alternative policies, technical strategies and solutions, and socio–economic–environmental impacts. These capabilities will support entities and programs engaged in the development, acquisition, and implementation of technologies and systems.

The comprehensive description of the system and its complexities emphasized in this application is an important foundation for the planning and decision actions of the multiple stakeholders. Decision methods and theories described in other case studies can be drawn upon as applicable to their specific roles and responsibilities.

13.3.2 Multiscale Modeling of the Healthcare System

Analyses of the myriad social, policy, economic, and technical aspects of the healthcare system can be supported using models applied by Health and Human Services (HHS)/Centers for Medicare and Medicaid Services (CMS) and VA for:

- Long-term economic forecasts, HHS-CMS (Boards of Trustees 2011)
- Budget projections, VA Enrollee Health Care Projection Model (EHCPM) (Harris, Galasso, and Eibner 2008)
- Program planning, HHS (Lewin for healthcare policy and stakeholder financial flows; VA EHCPM for program planning) (Lewin Group 2004)

Current models are either highly aggregated with little program detail, or when detailed are based on recent data and analysis for services and their costs as in the resource-based relative value scales (RBRVS) as a basis for Medicare reimbursements. They generally do not incorporate forward-looking technology, transformed services and processes, or outcomes that are important factors in all of the above applications.

Table 13.1 outlines the multiscale framework highlighting ESE models for the three principal scale-levels outlined in Figure 13.1, along with relevant healthcare data at each level for the programs and policy options they can address.

The ESE modeling application of major interest includes the planning, analysis, and evaluation of transformational healthcare programs. For this purpose, the multiscale characterizations described here are required to encompass critical attributes and their complex interrelationships.

Various organizations in the Department of Health and Human Services (2011a), other agencies, and universities maintain state-of-the-art information and analytical models related to healthcare as summarized below. These are very detailed and designed for specific purposes. Furthermore, recent trends in the release of government data are to provide raw data that may be structured as needed by users for their specific purposes, data of this form can be integrated to provide a modeling and analysis resource using ESE methods.

13.3.2.1 Descriptive Modeling and Analytic Methods

Specific models applied in this case study to Type 2 diabetes management are:

- RHSS and hybrid models
- Interindustry and macroeconomic model of the U.S. economy

13.3.2.1.1 The Reference Health Services System (RHSS)

Figure 13.2 describes the RHSS framework that positions the healthcare sectors of the economy in a sequence that delivers health services to an individual. When populated for a given base year and for projections to future years the framework provides a state-space representation of the expected or desired transformation. The darker backbone of the RHSS encompasses the delivery of direct health services by providers through the performance line of sight that couples programmatic and policy initiatives to defined outcomes. Other blocks outside the backbone denote supporting, enabling, and influencing activities.

Table 13.1   Multiscale ESE Framework for Healthcare

Multiscale Definitions	Healthcare Data and Analytics	Representation of Programs and Policies
1.a. Demographics (The public—including those in "patient" and "well" states) **1.b. The U.S. Economy**	Outcomes: Disease rates, morbidity and mortality Macroeconomic and interindustry models LIFT/ILIAD–97/376 sectors	Public health and wellness Income, jobs, and service demands Government programs and policies for VA and HHS/CMS Taxes, subsidies, and trust funds
2.a. Health Enterprise **2.b. Health Sectors** Providers and Facilities Technology Insurers, government and private	LIFT/ILIAD, 12 Health Sectors, BEA/Census Bureau—Census of Manufactures Data Lewin health benefits simulation model RHSS Framework	Services and policies for 12 healthcare sectors Provider incentives and disincentives Insurer and government payment mechanisms Pharmaceuticals and med. equipment
3.a. Healthcare System Services Delivery processes Resources **3.b. Workforce**	VA-EHCPM (enrollee health care projection model) RHSS health system model (reference health service system) demographics, health services	Investment in facilities and processes Technologies as a major influence on health services and cost Medical supplier's data and trends AHRQ technology assessments
Supporting Data: Outcomes Economics Resources Workforce	RBRVS, H-CUP, MEPS, CER (resource-based relative value scales) Comparative effectiveness CMS technology assessments	Technologies embedded in thousands of health and medical service Assessment for effectiveness

Source: F.L. Dougherty et al. 2012 © IEEE.

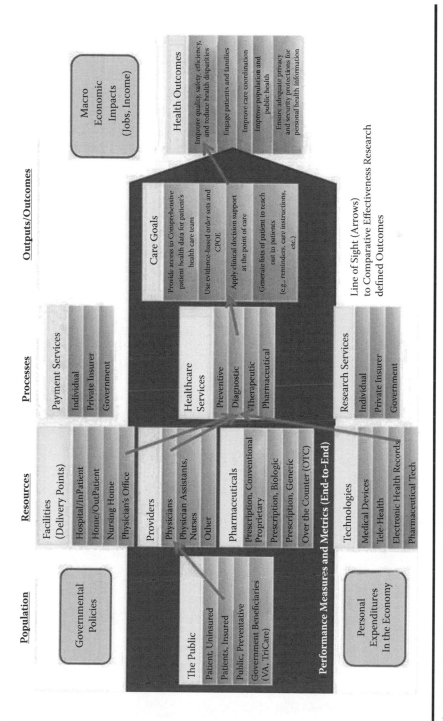

Figure 13.2   The Reference Health Services System (RHSS), a nested SoS. (Reprinted from F.L. Dougherty et al. 2012. © IEEE. With permission.)

There is an abundance of data available on the delivery of healthcare services and increased emphasis is placed going forward on the effectiveness of procedures and outcomes. The RHSS provides for the integration of these data elements as they become available in a comprehensive systems framework.

The application of the RHSS to a specific health service is outlined in Section 13.4 for Type 2 diabetes management using a hybrid modeling approach that describes the population at risk, the services provided, and the sequence of potential outcomes. The application demonstrates how the characteristics of specific services can be aggregated at the enterprise level and incorporated in models of the U.S. economy.

The RHSS can be formulated at the level of a practice, hospital, government provider, or national enterprise growing in complexity at these multiple scales of application. Data, by necessity, will be more aggregate at the higher levels. Note the positioning of the RHSS framework at the enterprise level in Figure 13.2 and at the healthcare system level (i.e., level 3) as a description of specific services delivered. As such, the RHSS provides the linkage between services system models and the enterprise level. The elements of the healthcare system and services in the RHSS can integrate the diverse datasets identified in Figure 13.2 to evaluate programs and policies, identify information gaps, and formulate research and programmatic priorities. These diverse datasets include the following information as summarized in Figure 13.2:

- Healthcare data at the system service level 3
- Resource-based relative value scales (AMA 2012)
- Patient demographics
- Healthcare data at the sector and economy levels 2 and 1, respectively
- Interindustry and econometrics relationships for personal income, GDP, and job impacts
- Long-term forecasts of the Medicaid/Medicare Trust Funds (Boards of Trustees 2011)
- Supporting HHS/CMS data sources
- Healthcare Cost and Utilization Project (H-CUP) (AHRQ 2011a)
- Medical Expenditures Panel Survey (MEPS) (AHRQ 2011b)
- Comparative effectiveness research (CER) (Department of Health and Human Services 2011b)
- CMS technology assessments
- Scope and content of major health models and data
- Health sector taxes and subsidies (current and proposed) (level 1)
- Long-term interindustry forecasting tool (LIFT), University of Maryland INFORUM Economic Model (INFORUM LIFT 2007) (levels 1 and 2)
- Lewin Health Benefits Simulation Model (Lewin Group 2004) (level 2)
- VA Enrollee Health Benefits Projection Model (VA-EHCPM) (Harris et al. 2008) (level 3)

Specific policy issues are mapped to the RHSS in Figure 13.3 to provide an influence diagram for a systems dynamics model as an element of a hybrid modeling approach.

13.3.2.1.2 Interindustry and Econometric Models of the U.S. Economy

These models encompass the entire healthcare system and represent its interrelationships with the other sectors of the U.S. economy. As an example of this class of models, the INFORUM long-term interindustry forecasting tool (LIFT) has been used to develop Medicare cost projections. The structure of LIFT encompassing the public, industry, and government is shown in Figure 13.4.

This class of models may be used to:

- Analyze the impact of policies and programs on GDP, sector dynamics, personal income, government costs and revenues, and employment.
- Examine the medical technology sector and supporting data (including older studies and proposals) and develop an approach to adding technology details.
- Develop a plan for integrating RHSS systems engineering methods.

There are nine health-related sectors defined in the INFORUM 97 sector LIFT dynamic interindustry model with three additions in the 376-sector Iliad static model. The elements of the healthcare enterprise defined in the RHSS framework are related to these sectors as follows to express all key stakeholder interests and interactions in policy analyses:

- Hospitals
- Hospitals, clinics, and associated practices (public and private, profit and nonprofit)
- Physicians
- Private sector practices; physicians, physicians' associates/assistants, nurse practitioners
- Nursing homes
- Other ambulatory healthcare services (Iliad)
- Home health services (Iliad)
- Social assistance, except child daycare services (Iliad)
- Dental, ophthalmic
- Dental offices
- Eye clinics
- Medical technology
- Hospital, office materials and supplies
- Diagnostic and therapeutic equipment

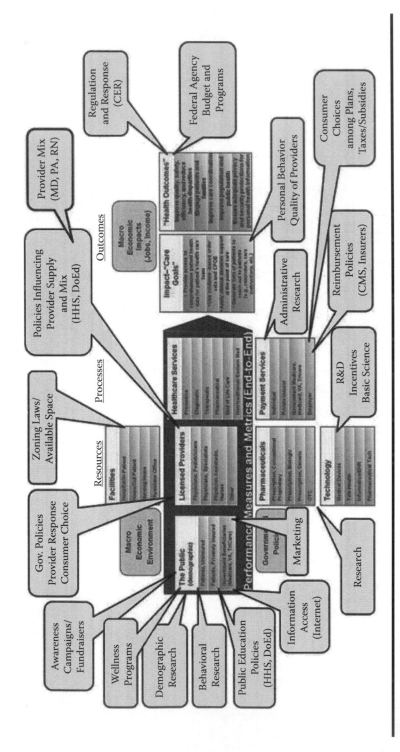

Figure 13.3 Strategies and policy influences mapped to the RHSS. (Reprinted from F.L. Dougherty et al. 2012. © IEEE. With permission.)

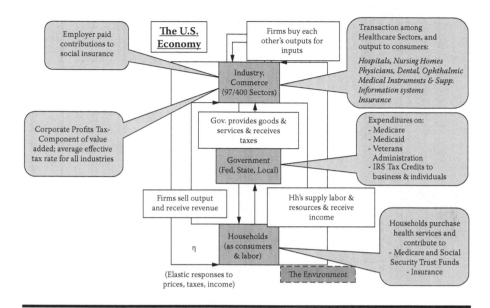

Figure 13.4 **Interindustry and macro model of the economy including the public, private sector, and government. (Reprinted from F.L. Dougherty et al. 2012. © IEEE. With permission.)**

- Laboratory services
- Lab materials and supplies
- Drugs
- Information systems
- Information technology
- Insurance
- Private plans
- Government
- Cooperatives
- Government institutions that deliver services (Veterans Administration, Department of Defense Military Health System)
- Government-financed programs (Medicare, Medicaid)

13.3.2.1.3 Visualizing the Transformation of Healthcare

The data and modeling methods outlined in this chapter are applicable to the planning and implementation of healthcare policies and programs supporting transition and development. Because of the scope and complexity of the system and its interrelationships, it is important to couple visualization methods with the models. Indeed, the RHSS provides a visual concept for the "supply chain" in delivering health services.

Geographical information systems (GIS) are commonly applied for automated mapping and visualization of data and analytical results. The *Dartmouth Atlas of Healthcare* (Dartmouth Institute for Health Policy and Clinical Practice 2010) is an excellent example of this capability. Most commercial products incorporate direct linkages to relational and object-oriented databases describing:

■ Demographics and resources in place
■ Facility configuration and capacities
■ Transformation program and project status

If the methods developed for planning and implementation of healthcare policies and programs and infrastructure projects follow the recommended approach, they can be supplemented to provide tools for operational management by public or private sector stakeholders.

13.4 Illustrative Healthcare Dynamics Analyses[*]

The multiscale approach using the RHSS framework and hybrid models has been applied to a specific health service: Type 2 diabetes management. This condition was chosen because of the significant health impacts and growing costs of diabetes. The progression of unmanaged diabetes to serious medical conditions including circulatory and cardiac conditions is modeled. The impacts of policies and new technologies are analyzed using a hybrid modeling approach including linkages of the following methods to analyze this nested SoS:

■ Agent-based methods describing the population to be treated
■ Systems dynamics methods to represent the influence of policies and technologies on health service delivery
■ Discrete event state-space descriptions of the progression of the disease with and without management
■ Interindustry and macroeconomic models to capture the impacts on the healthcare sector in the context of the U.S. economy

13.4.1 Healthcare System and Service Transformation and Technology Assessment (Microscale)

Selected elements of the systems engineering methods described are applicable to major healthcare policy initiatives at micro and macro levels. Healthcare policy is focusing on health informatics technologies to control costs, reduce medical errors, and extend high-quality coverage to the underserved population. Informatics technologies of particular interest include:

---

[*] Dougherty, F.L. et al. 2012. © 2012 IEEE. Modeling the U.S. healthcare system as an enterprise—multiscale hybrid data analytic methods. In IEEE International Systems Engineering Conference, Vancouver, B.C., March. With permission.

- Electronic Health Record (EHR)
- Computerized Physician Order Entry (CPOE)
- Decision Support Systems (DSS)
- Health information networks, sensors, and telemedicine

Introduction of these technologies involves significant process change in the way that health services are provided. Regulatory and cost reimbursement policies have a major impact on their introduction and effectiveness. This chapter outlines a systems engineering methodology for planning and analysis of such policies and the associated process changes. The systems engineering method may be coupled with economic models that describe the evolution of the health sector within the overall U.S. economy, now at 18% of GDP, to improve the utility of both methods for planning and analysis of programs and policies.

To perform technology assessments and to analyze policy impacts, the data and assumptions discussed above and RBRVS cost data can be applied at a microscale to a specific patient–process thread in the RHSS as shown on Figure 13.5, in this case for Type 2 diabetes management.

Input parameters to the patient-centric thread include:

- Patient health status, demographics
- Specific diagnosis, treatment, follow-up processes, with service modifications to be analyzed
- Informatics technologies employed
- Reimbursement and regulatory policies
- Resources required for each activity, for example, location of the services (in-patient, out-patient, day clinic, in-home monitoring with sensors, etc., that affect facility and overhead support costs)

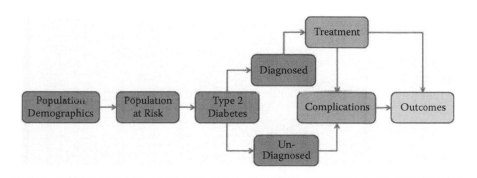

Figure 13.5 Hybrid model of healthcare delivery for diabetes management and treatments, a process thread extracted from the RHSS. (Reprinted from F.L. Dougherty et al. 2012. © IEEE. With permission.)

Model results that can be obtained include:

■ Cost of health services
■ Patient time required
■ Utilization of resources
■ Resource-based relative value scale

This form of activity-based healthcare process model can be applied to groups of patients at the regional level and in more aggregate forms at the national level using representative conditions and processes.

The analysis of the diabetes management example presented above describes the potential cost impacts of the selected policies and programs. These impacts alter the resources applied to this health service and affect the multiple health sectors that make up the U.S. economy. Demonstration results obtained by examining the diabetes thread from the RHSS draw on published estimates of specific costs and benefits and are shown in Figure 13.6. The baseline projection without Telecare was developed using a microsimulation process model; we combined Agency for Healthcare Research and Quality (AHRQ) MEPS survey data (AHRQ 2011b) for the costs of treating diabetes-related diagnostic codes and integrated them with census projections for changing demographics. There is a wide range of published estimates for near-term transformation costs and long-term benefits of the EHR and telemedicine; the results shown here are based on mid-range technology cost and outcome estimates. These effects influence the growth of the health sectors and

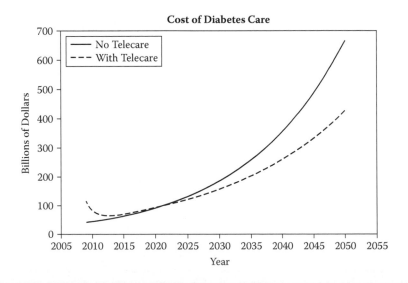

Figure 13.6 **Projected cost impacts for policy interventions to improve Type 2 diabetes management. (Reprinted from F.L. Dougherty et al. 2012. © IEEE. With permission.)**

can be analyzed by linking the RHSS to an interindustry macroeconomic model such as LIFT. The approach to this linkage is described in the following section.

13.4.2 Coupling with an Interindustry Model of the U.S. Economy (Coupling Micro- and Macroscales)

The approach to coupling a process model at the microscale to a model of the U.S. economy was applied by Steckley et al. (2011) for the energy sector of the economy to analyze the interindustry and macroeconomic impacts of transforming the energy system to a higher level of efficiency through tax and subsidy policies. The demographic and health service attributes of the micro application above for Type 2 diabetes management may be mapped to the sectors of the economy in the LIFT model using this approach.

Figure 13.7 illustrates the approach to coupling the micro- and macroscales. The RHSS microanalysis shows an important impact on health services, but only a modest impact in the context of the overall U.S. economy. A more comprehensive analysis of policies and technologies across many health services to explore ways to moderate the rapid growth of healthcare costs while improving outcomes would exhibit a significant level of impact on the overall economy and is left for future applications.

13.4.3 Analysis of Healthcare Policy at the Macroscale

A policy application of the LIFT model using a simplified approach to the coupling described above is presented below. This application addresses the expansion of health services coverage in the uninsured population via a major expansion of the Medicaid program. This analysis demonstrates the application of economic interindustry modeling to a major national policy initiative.

The analysis performed and documented by Werling (2006) traces the industrial and economic impacts of extending health insurance coverage to an additional

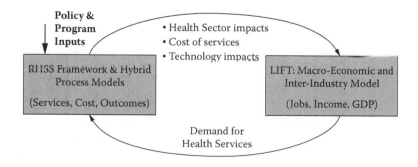

Figure 13.7 Coupling RHSS and hybrid models with interindustry macromodel of the economy. (Reprinted from F.L. Dougherty et al. 2012. © IEEE. With permission.)

20 million Americans over the next 10 years. It uses the INFORUM LIFT model of the U.S. economy to simulate an expansion of the existing Medicaid system to cover 9 million more children and 11 more million adults. The LIFT model is a 97-sector general equilibrium representation of the U.S. economy. Its detailed structure allows us to illustrate how health sector prices, incomes, and employment will respond to changes in insurance coverage and health expenditures.

To cover an additional 20 million enrollees, Medicaid expenditures are increased by $36.6 billion in 2007, an amount that rises to $70 billion by 2016. The analysis finds that a Medicare extension for these relatively young demographic groups results in a modest boost to the output, employment, and income of specific health industries. For example, compared to a baseline forecast, expenditure for hospital, physician, and other professional medical services expands by 1 to 3%. Overall nominal total national health expenditures are pushed up by only 1.3% in 2007 and to 1.7% by 2016.

From a macroeconomic perspective, the Medicaid expansion creates negligible changes. We do increase federal and state income taxes to fund the entire increase in spending. In nominal terms, collections rise by about 2.0 to 3.0% for federal taxes and 4.0 to 6.0% for state taxes. Nevertheless, compared to the baseline projection, overall real GDP falls by a trivial 0.02% by 2016 and the consumer price level increases by 0.17%. Indeed, the LIFT model shows that these enhancements to health expenditure can be fully accommodated by income tax increases and that output can be increased in the directly affected industries without appreciably affecting economic growth. The results are summarized in Figure 13.8.

Medicaid Population:	Total	Adults	Children	Aged	Disabled
Base Enrollment 2003, thousands	55,000	14,300	27,300	7,700	5,900
Expenditures per enrollee, $	4,200	1,900	1,500	7,200	17,400
Additional Enrollments, thousands	20,000	9,000	11,000		
Estimated Impacts:	2008	2010	2012	2014	2016
Nation. Health Exp/ (NHE), $Billion	2,400	2,700	3,100	3,600	4,200
Increase, $Billion	35	40	50	60	70
NHE, % of GDP	16.3	16.7	17.3	17.8	18.2
Increase, %	0.2	0.2	0.2	0.30	0.30
GDP Growth Reduction, annual %	0.01	0.01	0.01	0.02	0.02
Employment increase, k jobs		5	22	33	38
Fed & State Tax Increase, $ Billion		54	64	75	89

Above details of analysis (and more) are available for sectors: Personal Healthcare, Prescription Drugs/Medicines, Durable Medical Goods, Physician & Clinical Services, Dentists, Other Professionals, Hospitals, and Nursing Homes

Figure 13.8   Economic impacts of extending Medicaid coverage to one-half of the uninsured population. (Reprinted from J. Werling, Economic impacts of increasing Medicaid coverage. Inforum paper, September 24, 2006. With permission.)

13.5 Healthcare Enterprise Planning and Analysis Methodology

All stakeholders in the healthcare enterprise must plan and analyze policies and programs that affect their activities. These activities range widely in scope and scale and require multiscale data-analytic methods as described above. A planning, analysis, and evaluation methodology based on the use of the RHSS and the economic system model is outlined here in generic form. The methodology includes the following steps:

1. Define the scope and scale of potential impacts of the policy or program within the RHSS and health sectors.
2. Develop estimates of demands and levels of activities within the required scope for selected healthcare services.
3. Search and compile comprehensive healthcare information in the RHSS framework.
4. Select, populate with RHSS data, and apply applicable analytical models at the selected scale:
   ■ Service-level process models to develop cost and effectiveness estimates (RHSS and VA EHCPM-like enterprise model).
   ■ Agent-based models to characterize the patient population and characteristics, and provider contacts.
   ■ Systems dynamics models to define policy influences and complex interactions.
   ■ Economic forecasting models.
   ■ Resource allocation and optimization models.

The logic of this approach to sequential planning is clear; a key step is that of process definition and the identification of policy and program initiatives that will transform processes. Specific projects may be identified to remove barriers to the transformation. The impact of defined projects on service delivery and outcomes must be coupled to financial plans and implementation plans. Iteration is required back to the program definition step to add, delete, or redirect projects and to set priorities.

Risk is inherent in all healthcare programs and is a major factor that must be addressed in the planning and implementation phases, as well as in subsequent operations. The magnitude of risk is dependent on the scope of policies addressed, technologies to be employed (stable, emerging, new) and on the stability of the delivery systems. The major risk categories that are encountered are outlined below. Idealized conditions are used as examples to illustrate risk avoidance measures, but are never realized fully in practice.

13.5.1 Policy and Business Risk

Public policies generally reflect the public interest and preferences that balance economic impacts and sustainability of policies with health and safety objectives. Policies must provide a clear and stable business environment as a basis for investment. Private sector responsibilities in the industry–government partnership must be defined and understood by all parties for healthcare system activities.

The major risks encountered in national programs that seek to transform large sectors of the economy involve emergent behaviors in health services and supporting insurers in response to federal policies. These tend to dilute, delay, and even counter the desired objectives and must be subjected to critical analyses.

13.5.2 Program Risk

Program objectives must be clearly defined and planning must be integrated to ensure stable supplies of resources and labor to carry out the project and make a smooth transition to operations. Design and operating requirements must be stable, and constitute a major risk in this dynamic area.

The potential for schedule slippage exists on every program. The resulting program risk is much more than the simple sum of project schedule slippages as they are compounded by the cascading effect of interrelationships and interdependencies between related projects in other portions of the system. Contractors must have references of their experience in the projects to be undertaken.

13.5.3 Resource Risk

Physical resources (providers, facilities, and technologies) must be adequate to support the demand for healthcare services. Financial resources for new construction and operation, often by cash-strapped municipalities, must be available with stable long-term commitments. Trained human resources must be available, and training plans must be implemented to ensure an adequate labor supply.

13.5.4 Technical Risk

Proven processes and technologies must be employed to minimize technical risks. Component obsolescence and degradation can constrain system performance as technical components do not meet expectations or specifications. Many of the promising technologies involve information systems to maintain current medical records, reduce errors, and save money. There is much resistance to the use of such technology in clinical settings attributable to public acceptance, privacy, and implementation risks for these investments. The performance of individual projects is important; however, all related projects and activities must perform to meet overall system objectives of greater access at affordable cost.

Information systems used for monitoring and control of processes always present risks that merit special attention. Computers are available that are fault tolerant, and other backup measures may be taken to ensure the reliability of data and control applications. A reliable communication system must be available with provision for emergency power supplies during power outages. There should be prior experience with any software in similar applications and the software should be fully documented.

13.5.5 Health and Safety Risk

Many federal agencies are engaged in the regulation of safety, including the Food and Drug Administration (FDA), U.S. Department of Agriculture (USDA), Nuclear Regulatory Commission (NRC), and the Department of Transportation (DOT) National Transportation Safety Board (NTSB). Risks associated with this regulation include uncertain industry impacts and emergent behaviors in response to regulation. These risks are weighed against the benefits of regulation as part of the regulatory process. Again, dealing with uncertainties is a crucial step.

13.5.6 Emergent Behavior Risk

A major risk in healthcare modeling and analysis, or any modeling and analysis of complex systems, is the unforeseen emergent behaviors that inevitably arise and that can dominate as policies are applied and programs implemented. Such behaviors can be positive or negative in their impacts, and can sometimes be identified by drawing on the institutional expertise of managers in highly regulated environments and on experts in technology forecasting. Analytical formulations and results must be subject to such expert review. However, emergent behaviors are defined by some to be unforeseeable by any known methods; so in the end all policies and program plans must incorporate specific corrective capabilities to draw upon localized experiments and to deal with uncertainties and emergent outcomes. Our federal system indeed allows for pertinent testing and evaluation at the state and local levels; experience in states such as Massachusetts (2006) and Oregon (2010) is particularly noteworthy in this regard.

13.6 Governance

The importance of governance in major transformation programs is stressed by Hamaker (2003), and is especially critical given the interdependencies among public agencies at the federal and state levels and private institutions that make up the healthcare enterprise. Well-defined roles and responsibilities in this highly distributed and complex economic environment can facilitate transformation functions shared by these public and private entities through:

- Strong accountability for policy and systems elements of transformational policies and programs and support activities performed
- A virtual support structure with central mission integration, but distributed support resources and information
- The application of the mission-centric architecture as a transformation roadmap and as a template to determine responsibilities and manage accountability
- A comprehensive multiscale analytic capability to weigh costs and outcome benefits that inform managers, decision makers, and the public

Responsibilities are matched to the hierarchy of private sector and public entities in Table 13.2. The responsibilities of supporting agencies are noted. For the roles and functions indicated, the responsible entities are the principal managers and users; they are also accountable for their currency of plans.

The coordination of these governance elements does not come easily. All of the elements are the subject of detailed systems engineering analysis and design. Architectures can play an important role in governance as described here, but require specific additions to their scope and content as dictated by the specific roles and objectives. The RHSS can support the required coordination.

In the case of the healthcare system, the HHS has legislative responsibility to maintain current data and to develop projections of future healthcare demands and costs (for the Medicare "Trust Fund"). These information sources and projections are available to all parties. The private sector entities have their own information sources and outlook and utilize them along with government sources in their business and investment plans.

13.7 Conclusion

The complexities of healthcare policy and program formulation require robust analytic capabilities to address national, regional, provider, and public impacts. The RHSS and multiscale hybrid models can structure and integrate diverse information pertinent to all stakeholders to describe the physical and economic aspects of the U.S. healthcare system. This capability is increasingly relevant as many government data sources are increasingly in the form of a "fire-hose" of raw data that analysts must digest and structure for their specific purposes. Such information encompasses health sectors that deliver services and the supporting resource and financial forms. Data include human resources and facilities as well as organizational roles and responsibilities, business processes, enterprise resources, and information services and technologies needed to improve capabilities and to support this complex enterprise. As more effectiveness and outcomes-based data become available, these methods will support comprehensive planning, analysis, and evaluation of policies and programs.

Table 13.2   Summary of Governance Roles and Responsibilities for the Healthcare System

Governance Roles and Responsibilities: Public and Private Sector Entities		
National Policy	Policy Formulation Regulation Education	HHS/CMS Local and Federal Public interactions
Healthcare Services • Meeting Demand • Affordability • Health and Safety	Delivery Effectiveness Cost management Reimbursement Regulation Legal	Providers (Profit and Nonprofit) Government (VA, DoD, MHS) Government (HHS) State and Local Government
System Development and Operations	Hospitals, Clinic Ops Reporting Claims processing	Industry Government
Services and Components of Systems, Technology	Diagnostic Laboratory Pharma Home sensors and equipment	Industry
Research	R&D incentives Longer-term basic research Pharma strategies	Industry NIH Government–industry collaboration
Data and Information	Privacy Security Interoperability	Government, HHS Industry
Technology Selection	Diagnostic Therapeutic Preventive	HHS-CMS, FDA Providers

No single modeling method can address the range of policies and programs under consideration to transform the U.S. healthcare enterprise. The multiscale concepts described in this chapter can be tailored and applied to this wide range of analytical challenges.

The micro- and macroapplications demonstrate the range of analyses that can be performed with coupled systems engineering and economic models that are described in this methodology chapter. The initial efforts are described and work is in progress on more complete alignment of the modeling methods. The major features of the combined systems engineering–economic modeling approach include:

- Moves beyond, but draws upon, piecemeal/static analyses
- Demonstrates feasibility of a comprehensive transformation analytics framework
- Can address a complex "challenge analytics case" with combined, interactive, and phased strategy elements, for example:
  - Analyze impacts of Medicare Payment Advisory Commission (MedPac) recommendations (multiyear phase-in)
  - Incorporate Medicaid expansion case ($20M over five years)
  - Blend HCTC expansion with subsidized COBRA (Consolidated Omnibus Budget Reconciliation Act; five-year phase-in)
  - Cooperative plan or public plan (five years)
  - Selective technology change (EHR, telemedicine, biologics, etc.)
  - Tax employer-provided benefits (sensitivity to various levels)
  - Delphi analysis of outcomes
  - Exposes key assumptions, complex interrelationships, and analytic results in dashboard for interactive stakeholder analysis
  - Can analyze key POET factors and impacts of "pay for performance" policy for Medicare
  - Draws upon current piecemeal/static analyses and data
  - MedPac Medicare cost management, and the like
  - Outcomes and comparative effectiveness research (CER) data (emergent)

There are also major uncertainties and data gaps in the representation of healthcare service activities and processes related primarily to effectiveness and cost, but including incentives and disincentives to stakeholders in government policies. This behavioral information is very important and cannot be ignored in policy analysis. It is far preferable to develop even rough estimates of such critical factors and subject them to sensitivity analysis than to ignore them in the interest in constructing "validated" models.

The development and maintenance of health is essential to our society. Successful market economies rely on a free flow of high-quality information on local resource markets and processing capabilities and inventories. Much of the current emphasis is on financial market data; physical data on services delivered and suppressed demand are less accessible indicating the following needs:

- National accounts and statistics should address requirements for health and safety data from private sector enterprises and responsible government agencies.
- An integrated planning methodology must be employed to ensure that projects are viable in the context of the overall system, and that all potential risks are identified.
- The systems methodology can reveal inconsistencies and imbalances in market data that must be understood by all participants in a project.
- Uniform approaches to effectiveness and cost accounting should be developed and promulgated among stakeholders.

Projects and policies often fail due to deficiencies in the infrastructure. A large number of these must be considered to be "system failures" and indicate that the system configuration, capacities, and bottlenecks must be defined and managed by private and public sector institutions. In addition:

- The role of stakeholders must be defined.
- All critical labor, material, and resource requirements for healthcare transformation should be identified and their markets should be tracked.
- Government policies affecting projects and private sector activity should be based on realistic physical and financial information. Policies and regulations should be based on a cost–benefit analysis taking technological capabilities and costs into account, and reflecting likely market responses.

Access and usability of the RHSS as an integrated healthcare information system can be promoted by the use of state-of-the-art information and communications technology. For example:

- Data should be organized in relational and object-oriented format for accessibility and ability to support ad hoc reports and analyses.
- A geographic interface should be provided for access and analysis across regions and demographic characteristics.
- System simulation or optimization with animation is feasible and should be used to help to make the information more useful for policy formulation and decision making.
- Material system models should be integrated with economic policy models.

It is clear that close and continuous coordination must be maintained between financial planning and policies addressing markets and the physical projects planned and implemented by the private sector and government. A comprehensive healthcare information system can support such coordination.

In summary, the concepts and supporting tools are sufficiently well developed for immediate application to a well-defined policy or program where the relevant technical and economic data can be collected from existing sources. Additional preparatory steps for a more comprehensive capability to address a broad range of program and policy applications include:

1. Map a more complete set of government and private sector data sources into the RHSS and health sectors in the models following earlier data architecture concepts.
2. Expand the capability to derive more specific data and information for:
    a. Informatics and diagnostic/therapeutic technologies that can bend the curve in both directions, e.g., by disaggregating RBRVS data and reconstructing it for projections.
    b. CER results to help project future services and processes.
3. Draw on and apply technoeconomic analysis methods that have been proven in other domain applications.
4. Employ robust visualization methods to display and interpret complex regional and national impacts and trade-offs in a decision-support environment.

References

ACA (Affordable Care Act). 2010.

AHRQ, 2011a. Agency for Healthcare Research and Quality, *Healthcare Costs and Utilization Project*, http://www.hcup-us.ahrq.gov/.

AHRQ. 2011b. Agency for Healthcare Research and Quality, Medical Expenditures Panel Survey, http://meps.ahrq.gov/mepsweb/.

AMA. 2012. American Medical Association, Medicare RBRVS, *The Physicians Guide, 2012.*

Boards of Trustees Federal Hospital and Medical Insurance Trust Funds. 2011. *2011 Annual Report.*

Brookings Institution, Engelberg Center for Health Care Reform. 2009. *Bending the Curve, Effective Steps to Address Long-Term Health Care Spending Growth.* Report. Washington, DC.

Dartmouth Institute for Health Policy and Clinical Practice. 2010. *Dartmouth Atlas of Healthcare.* http://www.dartmouthatlas.org/.

Department of Health and Human Services. 2011a. *Directory of HHS Data Resources*, http://aspe.hhs.gov/datacncl/DataDir/index.shtml.

Department of Health and Human Services. 2011b. *Comparative Effectiveness Research Funding*, http://www.hhs.gov/recovery/programs/cer/

Dougherty, F.L., Padovano, C.A., Hoffman, K.C., and Thornton, P. 2006. Socio-economic impacts of health care informatics. Presented at American Public Health Association Annual Conference, Boston, November.

Dougherty F.L, Hoffman, K.C., Huntington, H., Jun, J., Klein, D., Le, K., Schoener, B., and Walters, M. 2012. Modeling the U.S healthcare system as an enterprise—multi-scale hybrid data analytic methods. In IEEE International Systems Engineering Conference, Vancouver, March.

Executive Order 13335. 2004. Executive Office of the President, Washington, DC, April 27.

Hamaker, S. 2003. Spotlight on governance. *Information Systems Control Journal,* 1. http://www.isaca.org/Journal/Past-Issues/2003/Volume-1/Documents/jpdf031-SpotlightonGovernance.pdf

Harris, K.M., Galasso, J.P., and Eibner, C. 2008. *Review and Evaluation of the VA Enrollee Health Care Projection Model.* Santa Monica, CA: Rand Corporation.

Health Technology to Enhance Quality Act of 2005. 2005. Proposed Legislation of the 109th Congress (S. 1262) (16 June).

INFORUM LIFT. 2007. Department of Economics, University of Maryland, Spring., http://inforum.umd.edu.

IRS (Internal Revenue Service). 2005. Health Coverage Tax Credit (HCTC) Overview. http://www.irs.gov/individuals/article/0,,id=109960,00.html.

Lewin Group. 2004. *The Cost of Tax Exempt Health Benefits in 2004.* Falls Church, VA.

Massachusetts Legislature. 2006. Health Care Access and Affordability Conference Committee Report (3 April). http://www.mass.gov/legis/summary.pdf.

NAS. 2005. National Academy of Sciences and Institute of Medicine, *Building a Better Healthcare System—A New Engineering/Healthcare Partnership,* Washington, DC: National Academies Press.

NYC (New York City) Department of Health. 2005. *Diabetes Prevention and Control.*

Oregon Health Care Policy and Research. 2010. http://www.oregon.gov/OHPPR/.

Rouse, W.B. 2008. Healthcare as a complex adaptive system. In *National Academy of Engineering—The Bridge* (Spring): 17–25. National Academy of Engineering, Washington, DC.

Steckley, S.G. et al. 2011. Energy demand analytics using coupled technological and economic models. *The Energy Journal* (October). 32 (Special Issue) http://www.iaee.org/en/publications/ejarticle.aspx?id=2460.

Werling, J. 2006. Economic impacts of increasing Medicaid coverage. Inforum Paper (24 September): 1–10.

Epilogue

Enterprise Systems Engineering and Architecting— Lessons Learned and the Road Ahead

Kenneth C. Hoffman, William J. Bunting, Anne Cady, Christopher G. Glazner, and Leonard A. Wojcik

Understanding the dynamic behaviors of the enterprise is essential to planning and managing the acquisition and implementation of transformational systems and technologies to improve mission and business performance. The framework and methods in Section I describe the scope and content of this emerging discipline of enterprise dynamics and the significant role it plays in advancing systems engineering to the enterprise level (enterprise system engineering and architecting, ESE/A). Emphasis is placed on overcoming past deficiencies in static and deterministic enterprise methods to enable the agile enterprise. This is accomplished in the ESE/A process as outlined in Section I and applied to case studies in Section II.

ESE/A is a useful process as now applied and is evolving at a rapid rate. The process deals effectively with the policy, organization, economic, and technology elements of the enterprise as demonstrated in the case studies. Perhaps the most difficult challenge is the organizational aspect, as this element tends to be more static than the ever-changing policy, economic, and technology dimensions. Further work is required on enterprise dynamics methods to deal with the complexities and rapid pace of societal change.

The unifying concept described in Chapter 4 is one view of that road ahead based on a timelier state-space description of the enterprise achievable through tighter coupling with management information systems such as enterprise resource planning. The definition of the core control–influence–uncontrolled (C-I-U) state-space model at the enterprise operational scale, and the definition of dynamic sense-and-respond mechanisms, provide unifying capabilities directed both at the higher scale of the economic sector in which the enterprise is embedded, and the lower organizational scales. Greater fidelity may be added to the core C-I-U model as required for analysis as the source modules are fully developed to provide more detailed and complete state-space representations. Methods of tagging the state-space descriptors to identify their temporal position in the life cycle, modular size, scale, and other perspectives should be investigated further, for example, semantic technology.

Improved methods to address uncertainties and to incorporate probabilistic factors in the framework are needed for decision support, coupled with visualization and gaming capabilities. The external environment is perhaps the most difficult challenge in this regard. More effective environmental scans are essential and must be coupled with more advanced game-theoretical and "what if" analytics.

Several of the methods represented in the preceding chapters address dynamic interactions of the enterprise with its external, mostly uncontrolled, environment, for example, control-theoretic, game-theoretic, highly optimized tolerance, system dynamics, and agent-based methods. These are employed in source modules characterizing specific functional areas of the enterprise. Future research will explore source modules describing competitive, adversarial, and deceptive behaviors as important dynamic elements.

The models discussed in previous chapters can be applied to "play out" the assumptions that enterprise decision makers have about the future, and can thereby be used for qualitative "what if" analysis of different acquisition strategies and management approaches. The models have been validated qualitatively based on extensive and diverse programmatic and operational management experience. It is expected that the models would need extensive calibration to be useful for quantitative prediction in specific enterprise systems engineering applications. Calibration, validation, and application may be best accomplished for specific agency planning and program management use through an integrated gaming facility supported by a diverse set of descriptive, prescriptive, and predictive models with a core state-space representation.

The analyses using multiple models described above indicate several planning and management guidelines for the acquisition of new capabilities in an enterprise system of systems. The models highlight important aspects to be addressed in the business case prepared prior to the acquisition. These span the full life cycle of acquisition and operations with attention to the "one-third rule" (the optimum acquisition time as one-third of total system lifetime) as well as other variable factors to ensure the targeted benefits of the transformation program. Significant delays require major updates to the business case and strategy to estimate a revised operational lifetime and the resultant benefits. The definition of maintenance and ageing effects over the system of systems' lifetime, and the sensitivity of the business case to those factors must be given more attention than is the usual practice.

Major uncertainties in the internal and external environments must be analyzed and estimated. Although the current formulation of the model is based on deterministic external impacts that are postulated in advance, it is generally possible to estimate ranges of periodicity and the magnitude of such impacts *a priori* based on the complexity of the system and environment and experience. Sensitivity analysis of the effect of such program impacts can and should be performed as an important part of the business case.

Significant effort must be devoted to characterizing potential emergent behaviors in a complex enterprise system of systems. Results of agent-based modeling and other methods dealing with such behaviors can be incorporated in the optimal control model, a subject of ongoing research directed at unifying multidisciplinary modeling perspectives with a more detailed set of enterprise state-space characteristics in a piecewise linear optimal control theory representation.

The analyses and guidelines bear directly on appropriate transformation and acquisition strategies. Where significant uncertainties and persistent events ("hits") are anticipated that interfere with a program, analytical results indicate the benefits of an incremental transformation or acquisition strategy that is adaptive to those impacts. This can be accomplished with modular development and commercial off-the-shelf integration strategies, with a managed services acquisition strategy drawing on proven commercial services, or a combined approach. In any case, the acquisition strategy must be based on program complexities and risks, business objectives, and environmental considerations such as represented in this optimal control model of the enterprise.

We note that the simple optimal control model of the system of systems engineering (SoSE) process generates a "one-third rule," which can be used as a baseline for comparison with real programs. We generalized the base model by including a specific parameter interpreted as "internal complexity." The model predicts that engineering efforts with very low internal complexity, that is, with few interacting parts, are optimally completed in times that are short relative to total lifetime. Furthermore, we consider operational ageing effects; these tend to produce an optimum acquisition time, which is less than that for a system without such ageing effects. Finally, we consider the whole C-I-U range with limits on control variables and external complexity, modeled as exogenous hits to the enterprise transformation effort (as a SoSE process). Such external complexity can produce a variety of optimal responses, depending on the details of the uncertainty regime, including rapid agile acquisition, or very gradual acquisition with extensive prototyping.

Results from optimal control models suggest that decision making about systems acquisition and transformation management should explicitly take into account expected system lifetimes, and the factors modeled here: internal and external complexity, anticipated internal and external impacts, as well as operational ageing effects. Considering enterprise complexities in this context, the general guidelines suggested in this chapter can help achieve better return on investment. However, this research is still at an early stage, and should be integrated with other modeling approaches to provide a full picture for enterprise decision making. In particular, we recommend further research on probabilistic modeling of the dynamic processes to help account for uncertainties.

The value of dynamic enterprise modeling and simulation and quantitative methods is clear. Quantification is important where feasible and where data can be estimated. Also, the qualitative behavior of quantified models may be an excellent source of insight for complex systems engineering. Purely qualitative approaches are also appropriate for incorporation into a full suite of models.

Finally, this sourcebook represents a snapshot of the state of an enterprise dynamics discipline in its formative stages. The authors believe that the directions established can contribute to ever-increasing capabilities to deal with the complexities of enterprise transformation to achieve higher performance levels.

Looking ahead at emerging capabilities for dynamic analyses that will be available to managers, a new world of ubiquitous computing—predicted for the past three decades—is becoming a reality posing major management opportunities and challenges. Ubiquitous computing makes information available anywhere anytime on any device with powerful data mining and analytical capabilities. Synergistic forces between societal trends, such as consumerization, and emerging technologies that enable ubiquitous computing are fueling decentralized, collaborative, real-time decision-making and control. These changes challenge legacy concepts of organizational structures and require new thinking about how to model POET dimensions. Agile enterprises will evolve to maintain integrity and viability through continuous change and adaptation to increase operational performance and value to stakeholders.

Boundaries between enterprises and between their business systems and infrastructure will become fuzzy and fluid and enterprise ecosystems will proliferate, extending operations beyond traditional organizational boundaries. Customers and other stakeholders will be brought more directly into an "extended enterprise" in this new world of ubiquitous computing. Strategic outcomes—once the basis for enterprise planning and transformation—are becoming emergent rather than predictable and controllable. Evolutionary transformation and acquisition methods are needed to cope with the pace of change characteristic of agile, adaptive enterprises and may focus on capabilities and problem solutions rather than long-term implementation cycles for more comprehensive but less flexible systems. Perhaps the strongest counterforce that will modulate the impact of ubiquitous computing technologies on enterprise transformation is the complexity associated with implementing security and privacy controls required to avoid cyber catastrophes resulting from decentralized, mobile, and dynamically formulated business processes. Successful enterprises will navigate this space through evolutionary processes that exploit the significant opportunities and manage the inherent risks.

Acronyms[*]

AAS	Advanced Automation System
ABC	Activity-Based Costing
ABM	Activity-Based Management
AFSC	Air Force Systems Command
AHRQ	Agency for Healthcare Research and Quality
ANSI	American National Standards Institute
AOC	Air and Space Operations Center
ARRA	American Recovery and Reinvestment Act of 2009
B	Billion
BNL	Brookhaven National Laboratory
C2C	Command and Control Center
C3H	Cooperation, Collaboration, and Communications for Health
C3I	Command, Control, Communications, and Intelligence
CAASD	Center for Advanced Aviation System Development
CAS	Complex Adaptive Systems
CCG	MITRE Center for Connected Government
CCHIT	Certification Commission for Healthcare Information Technology
CEM	MITRE Center for Enterprise Modernization
CER	Comparative Effectiveness Research
CHIP	Children's Health Insurance Program
CIIS	Center for Integrated Intelligence Systems
CIO	Chief Information Officer
CIP	Critical Infrastructure Protection
CIU	Controlled, Influenced, Uncontrolled
CMS	Centers for Medicare and Medicaid Services
CONOPS	Concept of Operations
COO	Chief Operating Officer

---

[*] Specialized chapter-specific acronyms are not shown here and are identified in the text.

CPM	Critical Path Method
CTH	MITRE Center for Transforming Health
CTO	Chief Technology Officer
DHS	Department of Homeland Security
DoD	Department of Defense
DoE	Department of Energy
EA	Enterprise Architecture
EHR	Electronic Health Record
ERM	Enterprise Resource Management
ERP	Enterprise Resource Planning
ESE	Enterprise Systems Engineering
ESE/A	Enterprise Systems Engineering and Architecting
FAR	Federal Acquisition Regulation
FDA	Food and Drug Administration
FEA	Federal Enterprise Architecture
FERC	Federal Energy Regulatory Commission
FFRDC	Federally Funded Research and Development Center
FHA	Federal Health Architecture
FISMA	Federal Information Security Management Act
FLITE	Financial and Logistics Integrated Technology Enterprise
FTE	Full-Time Equivalent
FY	Fiscal Year
G&A	General and Administration
GAO	Government Accountability Office
GERA	Generalized Enterprise Reference Architecture
GIS	Geographic Information System
GNP	Gross National Product
GST	General Systems Theory
HCFA	Healthcare Financing Administration
HHS	Department of Health and Human Services
HMO	Health Maintenance Organization
HPH	Healthcare and Public Health
HR	Human Resources
HRSA	Health Resources and Services Agency
HS SEDI™	Homeland Security Systems Engineering and Development Institute
IEA	International Energy Agency
IEEE	Institute of Electrical and Electronics Engineers

IFAC	International Federation of Automatic Control
IPO	Integrated Program Office
IPRM	Integrated Project and Resource Management
IR&D	Independent Research and Development
IRS	Internal Revenue Service
ISO	Independent Systems Operators
IT	Information Technology
IV&V	Independent Verification and Validation
LAI	Lean Advancement Initiative
LoB	Line of Business
M	Millions
MEBN	Multientity Bayesian Networks
MEPS	Medical Expense and Performance Reporting System
MIT	Massachusetts Institute of Technology
MSR	MITRE Sponsored Research
NAE	National Academies of Engineering
NIH	National Institutes of Health
NIMH	National Institute of Mental Health
NIST	National Institute of Standards and Technology
NRC	National Research Council
O&M	Operations and Management
OMB	Office of Management and Budget
OR	Operations Research
ORMS	Operations Research and Management Science
PERT	Program Evaluation and Review Technique
PHR	Personal Health Record
PMA	President's Management Agenda
PMO	Program Management Office
POET	Policy, Organization, Economics, and Technology (as used here), also Policy, Operations, Economics, and Technology
PRM	Performance Reference Model
R&D	Research and Development
R2A2	Roles, Responsibilities, Accountability, and Authority
ROI	Return on Investment
RTO	Regional Transmission Organizations
S&A	Studies and Analysis
S&T	Science and Technology

SAGE	Semi-Automatic Ground Environment
SE	Systems Engineering
SEDI	Systems Engineering and Development Institute
SEPO	MITRE Systems Engineering Practice Office
SME	Subject Matter Expert
SoSE	System of Systems Engineering
SSA	Social Security Administration
T	Trillion
TEM	Technical Exchange Meeting
U.S.	United States
VA	Department of Veterans Affairs

Bibliography

Abbott, A. 1990. Conceptions of time and events in social science methods: Causal and narrative approaches, *Hist. Methods*, 23: 140–150.

ACA (Affordable Care Act). 2010.

AHRQ, Agency for Healthcare Research and Quality. 2011a. *Healthcare Costs and Utilization Project*, http://www.hcup-us.ahrq.gov/.

AHRQ, Agency for Healthcare Research and Quality. 2011b. *Medical Expenditures Panel Survey*. http://meps.ahrq.gov/mepsweb/.

Alderson, D.L. and Doyle, J.C. 2010. Contrasting views of complexity and their implications for network-centric infrastructures. *IEEE Transactions on Systems, Man and Cybernetics—Part A: Systems and Humans*, 40(4, July): 839–852.

Alderson, D.L., Li, L., Willinger, W., and Doyle, J. 2004. The role of design in the Internet and other complex engineering systems. Presented at the New England Complex Systems Institute (NECSI) International Conference on Complex Systems (ICCS), Boston.

AMA, American Medical Association, 2012. *Medicare RBRVS, The Physicians Guide, 2012*.

American Heritage Dictionary of the English Language, Fourth Edition. 2000. Boston: Houghton Mifflin.

Anderson, P., Meyer, A., et al. (1999). Introduction to the special issue: Applications of complexity theory to organizational science. *Organization Science,* 10(3): 233–236.

ANSI/IEEE 1471:2000—IEEE Recommended Practice for Architectural Description of Software-Intensive Systems.

Argyris, C. and Schön, D. (1978). *Organizational Learning: A Theory of Action Perspective*. Reading, MA: Addison-Wesley.

Aristotle. 1994. *The Complete Works of Aristotle: Revised Oxford Translation*, J. Boras (Ed.). Princeton, NJ: Princeton University Press.

Arthur, W.B. 2009. *The Nature of Technology: What It Is and How it Evolves*. New York: Free Press.

Assimakopoulos, V. 1992. Combining decision support tools and knowledge based approach for the development of an integrated system for regional energy planning. *Energy Systems and Policy*, 15(5): 245–255.

Axelrod, R. 1984. *The Evolution of Cooperation*. New York: BasicBooks.

Bar-Yam, Y. 2003. When systems engineering fails—Toward complex systems engineering. In *International Conference on Systems, Man & Cybernetics*. Vol. 2. Piscataway, NJ: IEEE Press, pp. 2021–2028.

Bernus, P. 2003. Enterprise models for enterprise architecture and ISO9000:2000. *Annual Reviews in Control,* 27: 211–220.

Bhagat, N.K. and Hoffman, K.C. 1980. *The Reference Material System—Materials Policy Information System for Renewable Resources, A Systematic Approach.* Saltillo, Mexico: Academic Press.

Boards of Trustees Federal Hospital and Medical Insurance Trust Funds. 2011. 2011 Annual Report.

Boiney, L.G. 2007. More than information overload: Supporting human attention allocation. Presented at 12th International Command and Control Research and Technology Symposium (CCRTS), Newport, RI, June 19–21. http://www.dodccrp.org/html3/events_12.html.

Boppana, K., Wang, Z., Wheeler, P., and Zborovskiy, M. 2005. AAS: Comparison of HOT model and system dynamics model. Briefing presented at Massachusetts Institute of Technology, December. Boston: Kluwer Academic.

Boppana, K., Wojcik, L., et al. 2006. Can models capture the complexity of the systems engineering process? Presented at the New England Complex Systems Institute (NECSI) International Conference on Complex Systems (ICCS2006), Boston, June 25–30.

Bousso, R. and Polchinski. J. 2004. The string theory landscape. *Scientific American* (September): 78–87.

Boutilier, C., Friedman, N., Goldszmidt, M., and Koller, D. 1996. Context-specific independence in Bayesian networks, uncertainty in artificial intelligence. In *Proceedings of the Twelfth Conference*, San Mateo, CA: Morgan Kaufmann.

Bozdogan, K. 2007. Enterprise architecture modeling, design and transformation: Defining the missing links. *Lean Aerospace Initiative*, Cambridge, MA: MIT, April.

Bozdogan, K., Glazner, C., Hoffman, K., and Sussman J. 2008. Computational Enterprise Modeling and Simulation. Lean Advancement Institute Annual Conference, Boston, MA, April 23.

Brookings Institution, Engelberg Center for Health Care Reform. 2009. *Bending the Curve, Effective Steps to Address Long-Term Health Care Spending Growth.*

Brouwer, F., Thomas, A.J., and Chadwick, M.J. (Eds.) 1991. *Land Use Changes in Europe: Processes of Change, Environmental Transformations and Future Patterns.* Dordrecht: Kluwer Academic.

Brunner P.H. and Rechberger, H. 2004. *Practical Handbook of Material Flow Analysis.* Boca Raton, FL: CRC Press.

Bryson, A.E. and Ho, Y.-C. 1975. *Applied Optimal Control.* Washington, DC: Hemisphere.

Bunting, W.J. 2009. (June). Reasoning on agency performance using line of sight evidential reasoning analysis. Paper presented at the Uncertainty in Artificial Intelligence UAI2009 Seventh Annual Workshop on Bayes Applications, Montreal, Canada.

Bunting, W.J. 2012. Reasoning on uncertain enterprise technology alignment for insight into attainment of enterprise transformation, *Journal of Enterprise Transformation*, 2(1, March): 50–79.

Burke, W.W. and Litwin, G.H. 1992. A causal model of organizational performance and change. *Journal of Management*, 18: 523–545.

Cantarella, G.E. and Sforza, A. 1991. Traffic Assignment. In M. Papageorgiou (Ed.), *Concise Encyclopedia of Traffic and Transportation Systems.* Oxford: Pergamon Press, pp. 513–520.

Carley, K.M. 1995. Computational and mathematical organization theory: Perspective and directions. *Computational and Mathematical Organization Theory*, 1(1): 39–56.

Carlock, P.G., Decker, S.C., and Fenton, R.E. 1999. Agency-level systems engineering for "systems of systems." *Systems and Information Technology Review Journal* (Spring/Summer): 99–110.

Carlson, J.M. and Doyle, J. 1999. Highly optimized tolerance: A mechanism for power laws in designed systems. *Physical Review E*, 60(2, August): 1412–1427.

Caro, R.A. 1975. *The Power Broker: Robert Moses and the Fall of New York*. New York: Vintage.

Chung, L., Nixon, B.A., Yu, E., and Mylopoulos, J. 1999. Non-functional requirements in software engineering. Boston: Kluwer Academic.

CIO (Chief Information Officer) Council. 2001(February). *A Practical Guide to Federal Enterprise Architecture.*

Clark, T. and Jones, R. 1999. Organisational interoperability maturity model for C2. In *Proceedings of the International Symposium on Modeling and Analysis of Command and Control.* Research and Technology Symposium, United States Naval War College, Newport, RI, June 29–July 1.

CORRIM (Committee on Renewable Resources for Industrial Materials), National Research Council. 1976. *Report of the Committee on Renewable Resources for Industrial Materials.* Washington, DC: The National Academy of Sciences.

Coyle, R.G. 1996. *System Dynamics Modelling: A Practical Approach.* London: Chapman and Hall/CRC Press London.

Coyle, R.G. 2004. *Practical Strategy: Structured Tools and Techniques.* Harlow, UK: Pearson Education, pp. 29–46.

Crutchfield, J.P. 2003. When evolution is revolution—Origins of innovation. In J.P. Crutchfield and P. Schuster (Eds.), *Evolutionary Dynamics: Exploring the Interplay of Selection, Accident, Neutrality, and Function.* Oxford: Oxford University Press.

Cyert, R. and March, J. 1963. *Behavioral Theory of the Firm.* Oxford: Blackwell.

Dahlgren, J. 2007. Real options and flexibility in organizational design. Presented at the 5th Annual Conference on Systems Engineering Research (CSER), International Council on Systems Engineering (INCOSE), Hoboken, NJ, March 14–16.

Dartmouth Institute for Health Policy and Clinical Practice. 2010. *Dartmouth Atlas of Healthcare.* http://www.dartmouthatlas.org/.

David, A.P. 2004. Probability, causality and the empirical world: A Bayes–de Finetti–Popper–Borel synthesis. *Statistical Science*, 19: 44–57.

Department of Defense Architecture Framework. 2009. DoD Architecture Framework V2.0, Volume 2: *Architectural Data and Models Architect's Guide.*

Department of Health and Human Services. 2011a. *Comparative Effectiveness Research Funding.* http://www.hhs.gov/recovery/programs/cer/.

Department of Health and Human Services. 2011b. *Directory of HHS Data Resources.* http://aspe.hhs.gov/datacncl/DataDir/index.shtml.

DeRosa, J.K. 2005. Thoughts on complex systems and enterprise systems engineering (ESE). Presentation to the Faculty of the College of Engineering and Mathematical Sciences at University of Vermont, Burlington.

Diez, F.J. and Druzdzel, M.J. 2007. Canonical probabilistic models for knowledge engineering (Technical Report CISIAD-06-01). Madrid, Spain: UNED.

DLA (Defense Logistics Agency). 2011b. *Strategic and Critical Materials Operations.* Report to Congress (January).

DoD (U.S. Department of Defense). 1997. *C4ISR Architecture Framework.* Vers. 2.

DoE (U.S. Department of Energy). 1998. Office of Environmental Management (June). *Accelerating Cleanup: Paths to Closure.* http://www.em.doe.gov:Publications:accpath.aspx.

DoE (U.S. Department of Energy). 2002. Office of Environmental Management. *Top-to-Bottom Review of the EM Program* (February 4). http://www.em.doe.gov/pdfs/16859ttbr.pdf.

DoE (U.S. Department of Energy) 2010a. Office of Environmental Management. *Roadmap for EM's Journey to Excellence* (16 December).

DoE (U.S. Department of Energy). 2010b. *Critical Materials Strategy* (December).

DoE (U.S. Department of Energy). 2011. *Draft Strategic Plan.* (February): 40–42.

Dooley, K.J. and van de Ven, A. 1999. Explaining complex organizational dynamics. *Organization Science,* 10(3): 358–372.

Dougherty F.L., Hoffman, K.C., Huntington, H., Jun, J., Klein, D., Le, K., Schoener, B., and Walters, M. 2012. Modeling the U.S. healthcare system as an enterprise—Multi-scale hybrid data analytic methods. In *IEEE International Systems Engineering Conference,* Vancouver, March.

Dougherty, F.L., Padovano, C.A., Hoffman, K.C., and Thornton, P. 2006. Socio-economic impacts of health care informatics. Presented at American Public Health Association Annual Conference, Boston, November.

EIA (Energy Information Administration). 2007. *US Energy Outlook.* U.S. Department of Energy.

Ellis, C.J. and van den Nouweland, A. 2000. A mechanism for inducing cooperation in non cooperative environments: Theory and applications. *Social Science Research Network* (February). http://papers.ssrn.com/sol3/papers.cfm?abstract_id=436522

Epstein, J. 2003. *Growing Adaptive Organizations: An Agent Based Computational Approach.* The Brookings Institution, Washington, DC.

Ethiraj, S.K. and Levinthal, D. 2004. Modularity and innovation in complex systems. *Management Science,* 50(2): 159–173.

Executive Office of the President, Office of Management and Budget. 2002. *The President's Management Agenda.*

FAA (Federal Aviation Administration). 2010. *FAA's NextGen Implementation Plan* (March). Washington, DC: Federal Aviation Administration.

Federal Enterprise Architecture Records Management. 2007. Federal Enterprise Architecture Program Management Office, Federal Enterprise Architecture Consolidated Reference Model Document Version 2.3.

Federal Segment Architecture Methodology. 2008. FSAM version 1.0.

Fischer-Kowalski, M. 1998. Society's metabolism. In: G. Redclift, and G. Woodgate (Eds.), *International Handbook of Environmental Sociology.* Cheltenham, UK: Edward Elgar.

Flint, D. 2005. The users view of why IT projects fail (February 4). Gartner, Inc. Research Paper, Washington, DC.

Forrester, J.W. 1961. *Industrial Dynamics.* Cambridge, MA: MIT Press.

Fowler, A. 2003. Systems modelling, simulation, and the dynamics of strategy. *Journal of Business Research,* 56(2): 135–144.

Fudenberg, D. and Tirole, J. 1991. *Game Theory.* Cambridge, MA: MIT Press.

Galbraith, J.R. 1973. *Designing complex organizations.* Reading, MA: Addison-Wesley.

Gass, S. 2002. Great Moments in HistORy, *OR/MS Today,* 29(5). http://www.lionhrtpub.com/orms/orms-10-02/frhistorysb1.html.

GERAM—Generalized Enterprise-Reference Architecture and Methodologies Appendix B ISO 15704 standard.

Glazner, C.G. 2009. Understanding enterprise behavior using hybrid simulation of enterprise architecture. Dissertation. Cambridge, MA: MIT.

Government Accountability Office. 2010. Organizational transformation, a framework for assessing and improving enterprise architecture management. GAO–10–8466, Washington, DC: U.S. Government Accountability Office.

Hacking, I. 2001. *An Introduction to Probability and Inductive Logic.* New York: Cambridge University Press.

Haefele, W. 1991. Energy, risk, and environment. *Energy Systems and Policy,* 14(4).

Hamaker, S. 2003. Spotlight on governance. *Information Systems Control Journal,* 1. http://www.isaca.org/Journal/Past-Issues/2003/Volume-1/Documents/jpdf031-SpotlightonGovernance.pdf

Hammer, M. and Champy, J. 1993. *Reengineering the Corporation: A Manifesto for Business Revolution.* New York: Harper Business.

Hardin, G. 1968. The tragedy of the commons. *Science,* 162: 1243–1248.

Harris, K.M., Galasso, J.P., and Eibner, C. 2008. *Review and Evaluation of the VA Enrollee Health Care Projection Model.* Santa Monica, CA: Rand Corporation.

Health Technology to Enhance Quality Act of 2005. 2005. Proposed Legislation of the 109th Congress (S. 1262) (June 16).

Hoffman, K.C. 1993. Management of enterprise-wide systems integration programs. *Journal of Systems Integration* (3/4, September): 201–224.

Hoffman, K.C. 1995. An integrated materials information system for infrastructure planning. *Journal of Systems Integration,* 5(2): 91–105.

Hoffman, K.C., 2011. Perspectives on the Validation of Energy System Models for Long-Term Projections, International Energy Agency ETSAP Workshop, Stanford University, July 9.

Hoffman, K.C. and Basile, P. 1982. Energy resources and environmental policy: Survey of analytical approaches. In *Proceedings of the First U.S.—China Conference on Energy, Resources, and Environment,* Beijing, November.

Hoffman K.C. and deTerra, N. 1984. Energy planning and long term strategies for developing countries. Presented at the *Fifth Annual International Meeting of the International Association of Energy Economists,* New Delhi, January.

Hoffman, K.C., Doernberg, A., and Hermelee, A. 1979 Impacts of new energy technologies as measured against reference energy systems. In *Proceedings of Conference on Non-Fossil Fuel and Non-Nuclear Fuel, Energy Strategies,* Honolulu (January). Elmsford, NY: Pergammon Press.

Hoffman, K.C. et al. 2007. Descriptive enterprise dynamics—A multi-disciplinary unifying framework. Presented at the Fifth Annual Conference on Systems Engineering Research (CSER), International Council on Systems Engineering (INCOSE), Hoboken, NJ.

Hoffman, K.C. and Jorgenson. D. 1977. Economic and technological models for evaluation of energy policies. *Bell Journal of Economics,* 8(2, Autumn): 444–466.

Hoffman, K.C. and Melancon, J. 1988. An information systems architecture for manufacturing/distribution enterprises. In *Proceedings of the ASME Manufacturing International 88,* Atlanta.

Hoffman, K.C., Pawlowski, T., Payne, D., and Zheng, K. 2005. Enterprise business, computing, and information services in a multi-agency environment: A case study in enterprise architect-engineering. In International Enterprise Distributed Object Computing Conference (EDOC) Workshop, IEEE Computer Society.

Huberman, B.A. and Glance, N.S. 1998. Fluctuating efforts and sustainable cooperation. In M.J. Prietula et al. (Eds.), *Simulating Organizations: Computational Models of Institutions and Groups.* Menlo Park, CA: American Association for Artificial Intelligence.

Hume, D. 2009. *A Treatise of Human Nature*—Volumes I and II. Frederick, MD: Merchant Books. (Original work published 1739).

INCOSE (International Council on Systems Engineering). 2010. http://www.incose.org/practice/whatissystemseng.aspx.

INFORUM LIFT. 2007. Department of Economics, University of Maryland, Spring, http://inforum.umd.edu.

IRS (Internal Revenue Service). 2005. Health Coverage Tax Credit (HCTC) Overview. http://www.irs.gov/individuals/article/0,,id=109960,00.html.

ISO 15704 2000. Industrial automation systems—Requirements for enterprise-reference architectures and methodologies.

Jensen, F.V. 2001. *Bayesian Networks and Decision Graphs*. New York: Springer-Verlag.

Johnson-Laird, P. 2009. Deductive reasoning. *WIREs Cognitive Science*, 1: 8–17.

Jorgenson, D. 1998. Economic and technological models for evaluation of energy policy. In *Growth*. Vol. I. Cambridge, MA: MIT Press, Chapter 9.

Josephson, J.R., and Josephson, S.G. 1996. *Abductive Inference: Computation, Philosophy, Technology*. Cambridge, UK: Cambridge University Press.

Juthe, A. 2005. Argument by analogy, argumentation. *Argumentation*, 19(1): 1–27.

Kalpic, B. and Bernus, P. 2002. Business process modelling in industry—The powerful tool in enterprise management. *Computers in Industry*, 47(3, March): 299–318.

Kavanagh, R. (Ed.) 1979. *Energy Systems Analysis*. Dordrecht: D. Reidel.

Kotnour, T. 2011. An emerging theory of enterprise transformation. *Journal of Enterprise Transformation*, 1: 48–70.

Kuras, M.L. and White, B.E. 2005. Engineering enterprises using complex-system engineering. In *INCOSE Proceedings*, Rochester, NY, July 10–15.

Kuras, M.L. and White, B.E. 2006. Complex systems engineering—Position paper: A regimen for CSE. Presented at Fourth Annual Conference on Systems Engineering Research (CSER), Los Angeles, April 7–8.

Laskey, K.B. 2008. MEBN: A language for first-order Bayesian knowledge bases. *Artificial Intelligence*, 172(2–3): 140–178.

Lee, T.H., Ball, B.C., and Tabors, R.D. 1990. *Energy Aftermath*. Cambridge, MA: Harvard Business School Press.

Levinthal, D.A. 2001. Dynamics of organizations. In *Modeling Adaptation on Rugged Landscapes*. Cambridge, MA: MIT Press, Chapter 11.

Levitt, R. 2004. Computation modeling of organizations comes of age. *Computational & Mathematical Organization Theory*, 10: 127–145.

Lewin Group. 2004. *The Cost of Tax Exempt Health Benefits in 2004*. Falls Church, VA.

Love, G. 2006. emRAM—Enterprise modernization assessment model, Briefing presented at MITRE, McLean, VA, October 20.

Lyneis, J.M. and Ford, D.N. 2007. System dynamics applied to project management: A survey, assessment, and directions for future research. *System Dynamics Review*, 23(2/3): 157–189.

Lyneis, J.M., Cooper, K.G., and Els, S.A. 2001. Strategic management of complex projects: A case study using system dynamics. *System Dynamics Review*, 17(3): 237–260.

Malone, T. 2004. *The Future of Work*. Cambridge, MA: MIT Press.

March, J. and Simon, H.A. 1958. *Organizations*. New York: Wiley.

Massachusetts Legislature. 2006. Health Care Access and Affordability Conference Committee Report (3 April). http://www.mass.gov/legis/summary.pdf.

Mathieu, J., Mahoney, P., and White, B. 2006. Agent-based acquisition stakeholder model. Briefing presented at MITRE, Bedford, MA.

Mathieu, J., James, J., Mahoney, P., Boiney, L., Hubbard, R., and White, B. 2007a. Hybrid system dynamic, Petri net, and agent-based modeling of the Air and Space Operations Center. In INCOSE Symposium, Systems Engineering: Key to Intelligent Enterprises, June 24–28, San Diego, CA.

Mathieu, J., Melhuish, J., James, J., Mahoney, P., Boiney, L., and White, B. 2007b. Multi-scale modeling of the Air and Space Operations Center. Presented at Symposium on Complex Systems Engineering, January 11–12. Santa Monica, CA: Rand Corporation. http:// cs.calstatela.edu/wiki/index.php/Symposium_on_Complex_Systems_Engineering.

McCarter, B.G. and White, B.E. 2007. Collaboration/cooperation in sharing and utilizing net-centric information. Presented at Conference on Systems Engineering Research (CSER), Stevens Institute of Technology, Hoboken, NJ, March 14–16.

Merriam-Webster On Line Dictionary. http://www.merriam-webster.com/dictionary/enterprise.

Merrifield, R., Calhoun, J., and Stevens, D. 2008. The next revolution in productivity. *Harvard Business Review*, June.

Milgrom, P. and Roberts, J. 1995. Complementarities and fit strategy, structure, and organizational change in manufacturing. *Journal of Accounting and Economics,* 19: 179–208.

Milly, P.C.D. et al. 2008. Stationarity is dead: Whither water management? *SCIENCE,* 319(1 February): 573.

Mingers, J. and Gill, A. 1997. *Multimethodology: Towards a theory and practice of combining management science methodologies.* Chichester, UK: John Wiley & Sons.

MITRE, C4ISR Architecture Working Group. 1998. *Levels of Information Systems Interoperability (LISI).* http://www.c3i.osd.mil/org/cio/i3/AWG_Digital_Library/index.htm.

Morecroft, J. 2007. *Strategic Modelling and Business Dynamics*. Chichester, UK: John Wiley & Sons, pp. 204–205.

NAPA (National Academy of Public Administration). 2007. Office of Environmental Management: *Managing America's Defense Nuclear Waste*. NAPA: 7-15 (December). http://www.napawash.org/wp-content/uploads/2007/07-15.pdf.

NAS, National Academy of Sciences and Institute of Medicine. 2005. *Building a Better Healthcare System, a New Engineering/Healthcare Partnership*. Washington, DC: National Academies Press.

NASCIO, National Association of State CIOs. 2002. *NASCIO Enterprise Architecture Development Toolkit*. Vers. 2.

Nash, J.F. 1950. Equilibrium points in n-person games. *Proceedings of the National Academy of Sciences,* 36(1): 48–49.

Nightingale, D. and Rhodes, D.H. 2004. Enterprise systems architecting: Emerging art and science within engineering systems. Presented at the Engineering Systems Symposium. Cambridge, MA.

NRC (National Research Council). 2006. Committee for Future Army Applications. *Network Science.* Washington, DC: National Academies Press.

NRC (National Research Council). 1976. *Renewable Resources for Industrial Uses.* Washington, DC: National Academy of Sciences.

NRC (National Research Council). 2005. *Network Science.* Washington, DC: National Academy of Sciences. http://darwin.nap.edu/books/0309100267/html/.

NYC (New York City) Department of Health. 2005. *Diabetes Prevention and Control.*

OECD (Organization for Economic Cooperation and Development). 2003. *Principles of Corporate Governance.*

OMB FEA (Office of Management and Budget Federal Enterprise Architecture). 2003. *Reference Models Series for Enterprise Architectures (Draft), Performance, Business, Systems and Components, Data and Information, Technology.*

OMB FEA. 2007. Consolidated Reference Model Document. Vers. 2.3.

Oregon Health Care Policy and Research. 2010. http://www.oregon.gov/OHPPR/.

Orton, J. and Weick, K. 1990. Loosely coupled systems: A reconceptualization. *Academy of Management Review,* 15(2): 203–223.

OnLine Star Inc. IET. 2007. Quiddity [Computer software]. Retrieved from http://www.iet.webfactional.com/quiddity.html.

Pearl, J. 1988. *Probabilistic Reasoning in Intelligent Systems.* San Francisco, CA: Morgan Kaufmann.

Pearl, J. 2000. *Causality: Models, Reasoning, and Inference.* Cambridge, UK: Cambridge University Press.

Player, S. and Keys, D.E. 1999. *Activity-Based Management.* 2nd ed. Hoboken, NJ: John Wiley & Sons.

Pollack, A. *NY Times.* 1993. Japan factory for making chip material destroyed. July 5.

Popper, S.L., Lempert, R.J., and Bankes, S.C. 2005. Shaping the future. *Scientific American* (April): 66–71.

Rabelo, L. and Speller, T.H., Jr. 2005. Sustaining growth in the modern enterprise: A case study. *Journal of Engineering and Technology Management,* 22: 274–290.

Rabelo, L., Eskandari, H., et al. 2007. Value chain analysis using hybrid simulation and AHP. *International Journal of Production Economics,* 105: 536–547.

Rabelo, L., Helal, M., Jones, A., and Min, H. 2005. Enterprise simulation: A hybrid system approach. *International Journal of Computer Integrated Manufacturing,* 18(6): 498–508.

Rahimzadegan, B., West, H., and Hoffman, K.C. 2006. HRW Cycles and lags model. Vers.2.0. Briefing presented at Cambridge, MA.

Richard, C., Beitel, J., Dunn, M.P., et al., 1998. AV-1998-113 Office of Inspector General Audit Report Advance(sic) Automation System, DOT: 8.

Rivkin, J.W. and Sigglekow, N. 2003. Balancing search and stability: Interdependencies among elements of organizational design. *Management Science,* 49(3): 290–311.

Rogich D. et al. 2008. Material flows in the United States: A physical accounting of the U.S. industrial economy. Report by the World Resources Institute (April).

Ross, J.W., Weil, H., et al. 2006. *Enterprise Architecture as Strategy.* Cambridge, MA: Harvard Business School Press.

Rother, M. and Shook, J. 2003. *Learning to See: Value Stream Mapping to Create Value and Eliminate Muda,* Cambridge, MA: Lean Enterprise Institute.

Rouse, W.B. 2005. A theory of enterprise transformation. *Systems Engineering,* 8: 279–295.

Rouse, W.B. 2008. Healthcare as a complex adaptive system. *National Academy of Engineering—The Bridge* (Spring): 17–25.

Schekkerman, J. 2006. *How to Survive in the Jungle of Enterprise Architecture Frameworks.* Victoria, BC: Trafford.

Schieritz, N. and Größler, A. 2003. Emergent structures in supply chains: A study integrating agent-based and system dynamics modeling. In *IEEE Computer Society: Proceedings of the 36th Hawaii International Conference on System Science.*

Schoener, B., Hoffman, K.C., Ibe, Z., and Steckler, S. 2008. Physical input-output model of biofuels production. MITRE Report, McLean, VA: MITRE.

Schum, D.A. 1994. *Evidential Foundations of Probabilistic Reasoning.* Hoboken, NJ: John Wiley & Sons.

Scott, W. and Davis, G. 2006. *Organizations and Organizing: Rational, Natural, and Open Systems Perspectives.* Upper Saddle River, NJ: Prentice Hall.

Senge, P.M. 1990. *The Fifth Discipline: The Art and Practice of the Learning Organization.* New York: Doubleday.

Sessions, R. 2007. *Building Distributed Applications—A Comparison of the Top Four Enterprise-Architecture Methodologies.* Microsoft Developer Network. http://msdn.microsoft.com/en-us/library/bb466232.aspx (Accessed April 17, 2012).

Siggelkow, N. and Levinthal, D.A. 2003. Temporarily divide to conquer: Centralized, decentralized, and reintegrated organizational approaches to exploration and adaptation. *Organization Science,* 14: 650–669.

Simon, H.A. 1962, The architecture of complexity, *Proceedings of the American Philosophical Society,* 106(6, Dec. 12): 467–482.

Simon, H.A. 1990. Prediction and prescription in systems modeling. *Operations Research,* 38(1): 7–14.

Snowden, D. and Stanbridge, P. 2005. The landscape of management: Creating the context for understanding social complexity. *E:CO* 6(1–2): 140–148.

Sousa, G.W.L., Van Aken, E.M., et al. 2002. Applying an enterprise engineering approach to engineering work: A focus on business process modeling. *Engineering Management Journal,* 14(3): 15–24.

Sowell, P.K. 2005. A readiness model for multi-agency interaction. MITRE Technical Report, McLean, VA: MITRE.

Sterman, J.D. 2000. Business Dynamics: Systems Thinking and Modeling for a Complex World. Boston: Irwin McGraw-Hill, p. 137.

Stevens, R. 2010. *Engineering Mega-Systems: The Challenge of Systems Engineering in the Information Age.* Boca Raton, FL: CRC Press.

Stubna, M.D. and Fowler, J. 2003. An application of the highly optimized tolerance model to electrical blackouts. *International Journal of Bifurcation and Chaos,* 13(1): 237–242.

Sussman, J.M. 2002. Representing the transportation/environmental system in Mexico City as a CLIOS. Presentation at the Fifth US-Mexico Workshop on Air Quality. Ixtapan de la Sal, Mexico.

Swinth, R.L. 1974. *Organizational Systems for Management: Designing, Planning, and Implementation.* Columbus, OH: Grid Inc.

Tatum, C.B. 1983. Decision-making in structuring construction project organizations. In Vol. 279 of *Technical Reports of the Department of Civil Engineering.* Stanford, CA: Stanford University.

The Open Group Architecture Framework. 2009. Version 9, *The Open Group.* Boston: The Open Group.

Thompson, J. 1967. *Organizations in Action: Social Science Bases of Administrative Theory.* New York: McGraw-Hill.

U.S. Congress, Office of Technology Assessment. 1982. *Review of the FAA 1982 National Airspace System Plan,* Library of Congress Catalog Number 82-600595, U.S. Government Printing Office, Washington, DC.

U.S. Congress. 2000. *Clinger Cohen Act* (PL 104-106).

VA Enrollee Health Benefits Projection Model (VA-EHCPM), 2010.

Van Lamsweerde, A. 2001. Goal oriented requirements engineering. A guided tour. In *Proceedings RE'01, Fifth IEEE International Symposium on Requirements Engineering.* Toronto: IEEE, pp. 249–263.

Venkateswaran, J. and Son, Y.-J. 2005. Hybrid system dynamic—Discrete event simulation-based architecture for hierarchical production planning. *International Journal of Production Research,* 20(15): 4397–4429.

Venkatraman, N. 1994. IT-enabled business transformation: From automation to business scope redefinition. *Sloan Management Review,* 35(2).

von Bertalanffy, L. 1956. An essay on the relativity of categories. *Philosophy of Science,* 22(4): 243–263.

Werling, J. 2006. Economic impacts of increasing medicaid coverage. Inforum Paper (September 24): 1–10.

White, B.E. 2005a. A complementary approach to enterprise systems engineering. Presented at the *National Defense Industrial Association Eighth Annual Systems Engineering Conference,* San Diego, October 24–27.

White, B.E. 2005b. Engineering enterprises using complex-system engineering (CSE). Presentation to First Annual System of Systems (SoS) Engineering Conference, Johnstown, PA, June 13–14.

White, B.E. 2006. On the pursuit of enterprise opportunities by systems engineering organizations. Presented at the IEEE Conference on Systems of Systems Engineering, Los Angeles, April 24–26.

Wiener, N. 1956. *I Am a Mathematician: The Later Life of a Prodigy.* Garden City, NY: Doubleday.

Wojcik., L.A. 2004. A highly-optimized tolerance (HOT) model of the large-scale systems engineering process. In *Student Papers: Complex Systems Summer School,* June 6– July 2, Santa Fe, NM: Santa Fe Institute.

Wojcik, L.A. and Hoffman, K.C. 2006. Systems of systems engineering in the enterprise context: A unifying framework for dynamics. Presented at 2006 IEEE/SMC International Conference on System of Systems Engineering (SoSE), April 24–26 (Digital Object Identifier 10.1109/SYSOSE.2006.1652268).

Wojcik, L.A. and Hoffman, K.C. 2007. Emergent enterprise dynamics in optimal control models of the system of systems engineering process. Presented at the 2007 IEEE/SMC International Conference on System of Systems Engineering (SoSE), April 16–18.

Yu, E.S.K. and Mylopoulos, J. 1998. Why goal-oriented requirements engineering. In E. Dubois, A. L. Opdahl, and K. Pohl (Eds.), *Proceedings of the Fourth International Workshop on Requirements Engineering: Foundations of Software Quality (REFSQ '98),* Pisa, Italy, June 8–9, pp. 15–22.

Zachman, J.A. 2008. *The Zachman Enterprise Framework.* Retrieved from http://zachmanframeworkassociates.com.

Zachman, J.A. 1987. A framework for information systems architecture, *IBM Systems Journal,* 26(3): 276–292.

Zhen, F. 1991. An SD model applicable to the study of rural energy development strategy in Beijing. London: Taylor and Francis; *Energy Systems and Policy,* 14: 213–226.

Zowghi, D. and Offen, R. 1997. A logical framework for modeling and reasoning about the evolution of requirements. *Proceedings of the Third IEEE International Symposium on Requirements Engineering,* Annapolis, MD.

Index

1/3 rule, 186–188

A

AAS, *see* Advanced automation system (AAS)
ABC, *see* Activity-based costing (ABC)
Abductive reasoning, 151–152
Abstraction for modeling, 114
Accountability between agencies, 86–87
Acquisition, definition, 65
Acquisition enterprise, 45
Acquisition failure rates, 46
Acquisition strategies, 327
Acronym glossary, 329–332
Activity-based costing (ABC) as tracking tool, 78
Activity-based healthcare process model, 312
Activity-based management (ABM)
    coupled with ERP, 9
    ESE/A and, 2
Activity specification, 86–87
Adaptive systems, 27
    modeling methodology, 42–43
Advanced automation system (AAS), 20; *see also* Federal Aviation Administration (FAA)
Adversarial strategies in operations research, 16
Agency-level planning
    extension for multiagency environment, 78
    resources, 87
Agent-based modeling, 27
    acquisition stakeholder model, 49
    case study, 212–233
    for dynamic elements, 42
    operator-environment model, 218–222
    strengths, 49
Agent measures of performance (MOP), 221–222

Agent response time, 229
Aggregated models lose detail, 303
Agile sense-and-respond enterprise, *see* Sense-and-respond enterprise
Air and Space Operations Center (AOC) case study, 215–233
Alignment of system aspects, 149–150
Alignment reasoning method, 148
Alternative architecture development, 139–142
Analysis tools
    application of, 41
    in case studies, 105
    dependent on management landscape, 42–43
    knockout analysis, 141
    multidisciplinary, 94–95
    sensitivity analysis, 141
    simulation modeling for, 129–130
Antisymmetry in LSERA, 162
AnyLogic® software, 126–127, 217
AOC, *see* Air and Space Operations Center (AOC)
Apprehension events workbench application, 262–263
Architectural hypothesis development, 123–125
Architecture
    alternative development of, 139–142
    connected to behaviors, 122–123
    as genomic code of enterprise, 78
    mission centric emphasis, 87
    proof of concept projection, 139–140
Architecture-based approach for method integration, 77
    in multiagency environment, 79
Argument model in LSERA, 52–53, 153, 154
Army systems and structures challenges, 10–11
Artifact entities in LSERA, 158

Printed and bound by CPI Group (UK) Ltd, Croydon, CR0 4YY

21/10/2024

01777085-0016